AF207380

# FALSE ALARM:
## THE RISE AND FALL OF THE CARBON DIOXIDE THEORY OF CLIMATE CHANGE

**Rex J. Fleming**

History Publishing Company LLC
Palisades, NY 10964

History Publishing Company LLC
PO Box700
Palisades, NY 10964


Ordering Information:
Quantity sales. Special discounts are available on quantity purchases by corporations, associations, and others. For details, contact the publisher at the address above.

Orders by U.S. trade bookstores and wholesalers. Please contact Big Distribution: Tel: (845) 359-1765; or visit www.historypublishingco.com .
History Publishing Company LLC PO Box 700
PO Box 700, Palisades, NY Palisades, NY10964
SAN 850-5942


Printed in the United States of America by
History Publishing Company LLC, Palisades, NY

# Preface

One of the purposes of this book is to help expose one of the longest scientific misrepresentations in modern history – that $CO_2$ causes *climate-change*. The climate on our planet is very complex. It begins with the fact that weather itself is chaotic and many scientific disciplines are involved in producing the 'climate'. The evolution of climate depends on how long a chain of weather events are assembled, and on what parameters are chosen from that ensemble of events for statistical evaluation.

There are many kinds of climate change that could be discussed. Climatic events cover a large spectrum of time and space scales. The goal of this multi-disciplinary science is to solve for the cause and effects of every specific space/time scenario – not necessarily choosing one over another. The effort required for a solution to *any one of these scenarios* is not trivial -- the required outcome involves quantifying the nonlinear effects associated with the different interwoven forces from several scientific fields.

This literary effort will present in layman's terms the proof that the primary definition of climate change, as viewed by the United Nations, and expressed as caused by carbon dioxide ($CO_2$) is a myth. Therefore the purported negative impacts on society by this cause are untrue – and that the continued promotion of this scientific fraud has caused financial losses to many, and loss of life for some of our world's poorest inhabitants.

The Intergovernmental Panel on Climate Change (IPCC) is a group established under the United Nations that has defined the length and domain of *climate–change as the multiyear change of the Earth's averaged global surface temperature*. This definition is consistent with the apparent goal of the organization – to ultimately eliminate the use of fossil fuels across the planet.

$CO_2$ is one of the most valuable molecules on Earth, but has been called out as the cause of the Modern Warming which began in 1850 at the end of the Little Ice Age. That this inappropriate statement has been perpetuated in virtually every country in the world, has been referred to a public relations campaign to lead the population to believe that global warming is man-made and a world crisis. In order to present a clear proof that this hypothesis is false and undeniably wrong, this definition of climate change must be followed here. However, that proof having been made, a broader perspective of climate-change will be presented.

The proof presented here is robust. It covers all three dimensions of possible concern. The first proof is the lack of correlation of $CO_2$ concentration with past *climate-change* regimes in the historical record – be they the intense Ice Ages of the past or in the intervening warm periods between the Ice Ages. There is a further 35 - year lack of correlation in 20th century records that exists within the current Modern Warming.

The second failure is the computer climate model results, which are the only evidence presented by the IPCC. This evidence may be exemplified as a lawyer making his case while standing in the middle of sinking sand. All the computer models have failed; they have over specified the degree of the current warming, and all past projections made from these models have produced warmings well above that which have occurred to date.

The third area of failure described is contained in the heart of this book which shows just why the climate models fail – $CO_2$ does not contribute any net heating to the atmospheric column – though both $CO_2$ and $H_2O$ contribute to a thermal blanket at the Earth's surface.

It will be demonstrated that three dynamic atmospheric processes routinely transfer surface heat upward through the atmosphere to a point where it becomes a trivial trace – ultimately radiated off to space. The data and programs that prove this result are accurately described herein and are available for anyone to reproduce. The three processes acting together also balance the outgoing radiation with the incoming solar radiation received from the Sun.

Arriving at the truth of nature's interactions is one of the goals of all science. However, in this case, the exposure and dismissal of this preposterous global warming cause -- has a much more important benefit – eliminating the needless pain inflicted on a great many citizens of the world.

The extra taxes collected on all aspects of fossil fuel production, delivery and use have been passed on by the companies that manufacture those products that have made our lives better. These extra taxes and costs have also made governments larger, not necessarily better, and have lowered the standard of living of virtually all of the world's inhabitants – except the rich.

These higher taxes and costs imposed on *domestic users* have a much greater negative impact. Rising fuel costs for transportation have caused riots in parts of Europe. Increased costs of power to our homes have grown significantly and affect: cooking the family food, heating homes, providing the power to run appliances and to maintaining sanitary conditions. These increased domestic taxes and higher energy costs do not affect the rich, but impact the middle class, and present an even hasher burden for the poor.

In the case of the continent of Africa there has been a deliberate effort to obstruct the use of fossil fuel. There are regions in Africa with a grossly inefficient energy system, in many cases no energy system, which has left tragic conditions: not enough clean water, nor enough food and few trees left -- women are cooking in tents burning dried dung as their heat source.

Data now 10 years old provided the following estimates: every day 30,000 people on this planet die of the diseases of poverty; one-third of the population on the planet do not have electricity; a billion people have no clean water; half a billion people going to bed hungry every night.

We as a civilization cannot sit and do nothing about these facts! A truly caring world of both liberals and conservatives would attempt to rectify these deplorable conditions as soon as possible.

There is potentially an even greater concern for all of humanity if the climate changes to an adverse cold period like that which has occurred in the not too distant past. The possibility of this event is discussed in the final Chapter. More $CO_2$, not less, will be required in colder times.

This book has been produced for the general public – for young adults and old. Hopefully it will both entertain and enlighten students, engineers, men and women of science, and policy makers. There has to be a coming together of all of us to produce a more humane civilization.

# Acknowledgments

There are many excellent scientists that have endured the wrath of those on the other side who believe that the current Modern Warming is caused primarily by carbon dioxide. This book is dedicated to those who have patiently put up with the misguided statements and actions from that segment of our society — including much of the media. One cannot acknowledge all, or even a significant portion of those individuals here. However, there are a few people that have directly contributed to this author's will to get the truth out for society. These are listed here with apologies to all the rest who have collectively contributed to these conclusions and deserve to be on this list, but cannot because of the space required.

The most important influence was the Chairman of the Atmospheric Science Department at the U. of Michigan, Prof. Aksel Wiin-Nielsen. Aksel, who has since passed away, knew the physics of the atmosphere as well as anyone. His very successful career was shortened, but not diminished, by disagreeing with the 'alarmist' view of climate change. Another important person, the hero of this book, is Prof. Richard Lindzen who was one of the first to challenge the science and the catastrophic views of the 'alarmist' side of the climate change argument. Other key people from a variety of backgrounds whose writings have been an influence include (in alphabetical order): Charles R. Anderson, Guus Berkhout, Freeman Dyson, Francois Gervais, Sherwood Idso, Patrick J. Michaels, Christopher Monckton, Patrick Moore, Nicola Scafetta and Fritz Vahrenholt. Many others could be on this list drawing from the names in the references.

Some of the important books that I have referred to that reflect the opposite position of the alarmist view of catastrophic climate change include those by the authors: Michael Crichton, James Delingpole, Mark Morano, Ian Plimer, S. Fred Singer and Dennis T. Avery, *O.G. Sorokhtin, G.V. Chilinger and L.F. Khilyuk,* Roy W. Spencer and *Henrik Svensmark and Nigel Calder.*

The successful effort to provide the registered letter to the United Nations, stating that there was no climate emergency, was an important international event. Those who provided the initiative for the effort, obtained the 14 signatures on the cover letter and solicited the signatures of the 500 scientists and professionals that signed the attachment – all deserve applause.

It has been a challenge to keep going with so much opposition, but the constant support by my lovely wife Kathy for the past 50 years, has sustained and encouraged me to complete the task.

# Table of Contents

Table of Contents

# List of Tables

# CHAPTER 1

# Introduction

This book will take the reader on a historical journey through time that begins with the creation of the universe, its evolution, the formation of our galaxy and later its solar system. Important details of Earth's early development and the emergence of our atmosphere will be revealed. This history is directly related to why "weather" in the atmosphere of planet Earth is so tremendously diverse. A clear distinction between weather variability and climate change is required. The actual story implied in the title begins around 1900 and ends some 120 years later.

Climate change itself is quite complex and covers a large spectrum of time and space scales. The goal of science is to solve the cause and effect of every specific space/time scenario – not necessarily choosing one over another, although practical considerations may dictate a certain sequence of scientific direction. The effort required for a solution to any one scenario is not trivial as evidenced by the interwoven scientific fields involved.

The world's atmosphere and oceans represent two turbulent fluids which interact with each other – the atmospheric winds drive the ocean currents and the oceans exchange heat and chemicals with the atmosphere. The atmosphere is chaotic, thus there are chaotic attributes within the ocean. The climate is constantly changing and one needs to define a specific period for climate change, and pick a particular area for that change as there are regional areas of change.

The United Nations' Intergovernmental Panel on Climate Change (IPCC) was established in 1988 by the World Meteorological Organization (WMO) and the United Nations Environmental Programme (UNEP). The IPCC definition of climate change represents *a very narrow view* and refers to a change in the state of the Earth's average *global surface temperature* over an extended period, typically decades or longer. It has also been revealed from within the UN that the purpose is much broader than *climate-change* – with ulterior political motivations.

The IPCC uses this particular *global* definition because they have attributed humankind as the primary cause of climate change by the use of fossil fuel and the subsequent release of carbon dioxide ($CO_2$) -- which has a global value over the planet of approximately 400 parts per million by volume (ppmv). The IPCC has a limited definition of climate change (just using surface temperature over the globe) as there are changes due to regional ocean/atmosphere dynamics, changes in ice volume concentrations, and other regional changes in various parameters. The main approach of the United Nations effort appears to be the use of scare tactics of impending disasters because of 'their expected extremely elevated global average surface temperature' in order to try and convince the world to eliminate or substantially reduce the fossil fuel industry.

Our purpose in this book is to expose one of the longest scientific misrepresentations in history – that $CO_2$ causes *climate-change* – again according to their definition. To that end, the IPCC definition of the length and domain of *climate–change* will be followed as the IPCC has defined them. It requires a significant external forces to cause such a *climate-change*. The exposure will begin with the introduction of this theory of the $CO_2$ cause of *climate-change* which began around the 1900 period. That discussion will occur in Chapter 6, but there is quite a bit of interesting history that has occurred since 1900 that must be revealed. The travelogue analogy will continue up to the present time and a bit beyond.

There are a number of scientific books that do not follow the thesis that $CO_2$ is a cause of our climate change. However, leaders of most governments and many others believe and support the notion that the Earth's atmosphere is a "greenhouse" with $CO_2$ as the primary "greenhouse" gas that is warming our planet. That this concept seems acceptable to some scientists and so many non-scientists is *perhaps* understandable. While there have been several periods of much *warmer atmospheric temperatures* than exist in our current time, the Modern Warming of the Earth's atmosphere began at the end of the Little Ice Age in 1850. The industrial revolution did not really take hold until the 1840 - 1870 period. It would be natural to believe that these two events, linked so closely in time, could be the reason for the rise in the current Modern Warming.

There is now a much clearer picture of an *alternative reason* for why the Earth's surface temperature has risen since 1850. A second scenario for this warming has emerged – a new theory of *climate-change* has matured over the past few decades. This not only explains the current Modern Warming, but also in whole or in part, many of the other major *climate-change* episodes that Earth has experienced. That new theory will be addressed and analyzed later in our journey.

The new theory was introduced by Svensmark and Friis-Christensen in 1997 (it is referenced later, and will be examined in detail at an appropriate period of our time travel). Briefly, when the Sun is "quiet", its magnetic field is weak, galactic cosmic rays are allowed entry to the Earth's atmosphere and produce vast areas of low level clouds that cool the Earth. When the Sun's magnetic field is strong, the cosmic rays are diverted, protected by that magnetic field, and do not penetrate our atmosphere and the Earth warms. The nature of cosmic rays, where they come from, and what they do to our atmosphere will be explained later – and in great detail in Appendix E.

Previous literary efforts have demonstrated the fallacies of the media scare tactics blaming $CO_2$ on a variety of events: more powerful storms, hurricanes, and extinction of species (some of these events are blatant falsehoods, some are real and important, but $CO_2$ is not their cause). No observational record has shown a clear correlation of $CO_2$ with *climate-change*! What has been missing is *a strong reason why*

*$CO_2$ has no impact on any net heat accumulation on a daily basis.* That reason requires a detailed calculation of many 1000's of $CO_2$ absorption coefficients subjected to the proper equations for radiative transfer. These calculations will be presented here – by two different, but closely related methods.

This book will provide a complete review of the role of $CO_2$ in the Earth's atmosphere. The logic of $CO_2$ involvement in changing the climate will be investigated from every perspective: reviewing the historical data record of Ice Ages with vast ice sheets, noting the interglacial periods of little or no ice, examining in further detail the 20th century data record, and evaluating the radiation role of $CO_2$ in the atmosphere. The radiation calculations, using the appropriate equations and data will be reviewed in great detail in Chapter 11.

The results of this review and examination reveal no role of $CO_2$ in any change of the Earth's climate – where *climate-change* is defined as stated above. The historical travel through time will reveal all one needs to know about this subject. Many different science disciplines will be visited along the journey. Each area of science will be introduced with a brief summary or an Appendix -- sufficient for virtually anyone to understand.

The reader will be introduced to the formation of our universe, its expansion into clusters of galaxies filled with vast numbers of stars, the formation of our Milky Way Galaxy, the creation of our Solar System, and the early evolution of Earth. This background will serve as a basis for the description of, and reason for, the *climate-changes* that have occurred on Earth over time.

Over the journey the reader will be introduced to a brief introduction into chemistry providing the reader with a simple interpretation of how isotopes are formed from the impact of cosmic rays; how the atmosphere produces such a tremendous variety of weather phenomena; and the principle components of the Sun. These elements include: the vertical structure, the solar dynamo that drives the Sun's magnetic field and a description of the solar wind. There is a discussion of the Sun's journey about the center of mass of the solar system.

The past one billion years have produced an atmosphere with a protective layer of a thermal blanket that changes with the seasons. There are three forces that: (1) create that thermal blanket, (2) cause the immediate transfer of heat upward -- thus powering the atmospheric circulation, driving the ocean currents, and implementing the irrigation system for planet Earth and (3) striving to maintain the required energy balance between the Sun and the Earth. These three forces are convection, latent heat release from evaporation, and the process of radiative transfer.

These three processes are described herein and then expressed in quantitative terms. There are occasional mathematical equations used to provide further clarity, but no mathematical manipulations are required of the reader to understand the concepts revealed.

The climate changes that have occurred over time are reviewed -- including those of the 20th century. This review will provide the reader with a deeper understanding of our world that will be both informative and hopefully awe-inspiring.

An important part of the book involves an appreciation of the value of $CO_2$. This "often demeaned molecule" is not the cause of climate change. Published research is presented on how enhanced $CO_2$ will provide greater security for mankind when the climate turns to a colder state.

The theory of *climate-change* now indicates a potential change from the current Modern Warming the Earth has experienced since the end of the Little Ice Age in 1850. That warming will go on as long as the dynamics of the Sun continue on their present course. However, the past *climate-change* cycles that have been uncovered from the variations of the Sun's magnetic field interacting with cosmic rays, reveals a potential significant change in the near future. That subject will be revealed in the last chapter of this book.

[Perhaps a little info about your guide on this tour would be useful -- my first paper published on the subject of climate was in

*Climate Dynamics* in 1993. I know both sides of the $CO_2$ issue having managed the NOAA Research Office funding scientists on the other side before leaving for Boulder CO to Manage the International TOGA project office. I was a Math major with a Physics minor at Creighton U, then later obtained my M.S. and Ph.D. in Atmospheric Science at the U. of Michigan. I have programmed a variety of atmospheric models: deterministic, Stochastic Dynamic Equations (SDE), and the combination SDE / Monte Carlo. {See my web page at http://rexfleming.com/}. I have retired three times and still self-fund my own research.]

# CHAPTER 2

# Creation of the Universe

The universe provides a wonderful vision to behold! How and when it was formed had been questioned by mankind through the centuries. However, it was not until the announcement in April of 1992 of the Cosmic Background Explorer (COBE) satellite results that firmly established that the universe began with the "big bang" -- an explosion that created matter, energy, space and time from an extremely small volume.

Many of the world's most famous astronomers made known their excitement; including Stephen Hawking, who was quoted[1] "It is the discovery of the century, if not of all time." A relatively brief background, may help explain the excitement of the COBE results. A history of the maturing universe and related facts supporting climate change are in Appendix A. Readers would do well to acquaint themselves with this material about this magnificent creation.

R. Tolman[2] in 1922 was the first to suggest that since the universe was expanding, it must be cooling from an initial very high temperature. G. Gamow[3] in 1946 suggested that only a rapid cooling from an initial high temperature could have produced the fusion of protons and neutrons required to produce the amounts of hydrogen and helium observed in the universe.

The Plank Satellite program of the European Space Agency was driven by those first COBE results, and later COBE results. That European program was a tremendous success[4]. The announcement of the results in 2015 – perhaps not as sensational as the 1992 announce-ment – but was perhaps even more important in confirming some loose ends in the astronomical world.

Further excitement was indeed created with the announced re-sults which should have been of special interest to all Earth's inhabit-ants. The age of the universe was 13.8 billion years since the "big bang". The composition of the universe was estimated to be 69% dark energy, 26% dark matter and 5% ordinary matter (e.g., mole-cules, planets and stars that we observe); and the expansion of the universe was confirmed to be *accelerating*[4].

Scientists suggested an extraordinary *inflation period in* the very early universe to explain the "horizon" problem which arose with the discovery of the cosmic background radiation being so homogeneous in all directions in space. Astronomers finding galaxies 10 billion light years away in two opposite directions implies that light (infor-mation) traveled 20 billion years but the universe is only 13.8 billion years old – so there must have been an extraordinary inflation period. Several models of inflation were proposed. In Guth's model[5] of the inflation it only lasted for a small fraction of a second (from $10^{-36}$ to $10^{-33}$ seconds and the once tiny region of space grew by $10^{60}$ times larger with essentially identical conditions throughout its volume[6].

The *timing of the growth of the galaxies throughout the universe* was extremely important for advanced life to exist and survive. There may be various forms of bacterial life somewhere in the universe, but for the development of mankind that has been achieved has required an amazing sequence of astronomical good fortune – or a benevolent external Force has had it all planned from the beginning[1].

The growth of galaxies began about 56,000 years after the Big Bang. As the universe cools from its very high temperature, more matter is being created by the high energy radiation. Through this expansion matter loses less energy than does the radiation. Eventually the energy density of matter (mostly in newly-formed nuclei) be-comes larger than the energy density of radiation (mostly massless

a gas because of the high temperature[6]. Nuclear reactions in the core consume hydrogen to produce helium and further heavier elements. Energy is released to the surface of the star as visible light.

Nuclear fusion begins when the temperature in the core is greater than 10 million degrees C. All stars generate their energy via the process of nuclear fusion. There is a large spectrum of star sizes. Stars change their luminosity and color with increasing mass size. Fig. 2.1 indicates the spectrum of star masses[9]. The mass of our Sun is considered to have a solar mass of 1.0.

Most stars have masses that are less than half the mass of our Sun. The observed peak in the spectrum is at about 0.1 solar masses. As the mass of stars increase, the rate at which fusion occurs increases – the more massive the star, the faster the hydrogen fuel is used up in the fusion process.

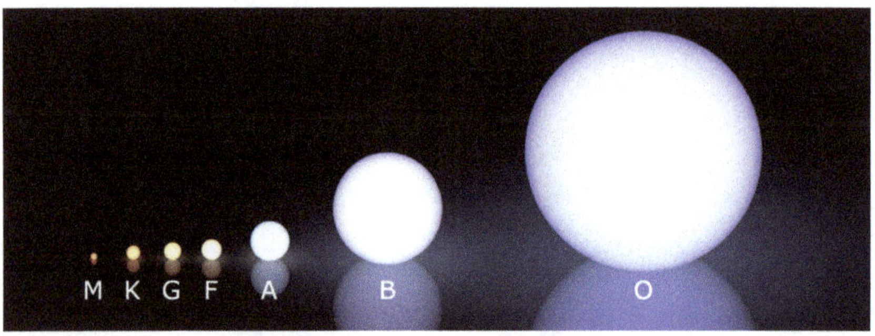


Fig. 2.1  Star types according to size – the G-type is the size of our Sun

The smaller M-type stars have *just enough mass to initiate nuclear fusion* and *could, in theory, have* lifetimes of several 100 billion years. If the star is small enough, heavier elements never reach the burning point and the fusion stops. The star stops producing energy and dies – shedding its outer layers with the light elements and perhaps a few heavier elements

The Sun in our solar system is a G-type star with a typical *yellow color*, which is discussed in far more detail in Chapter 12. It has a size that dictates a 10 billion year life – which is now half over. In 5 billion years the Sun will run out of hydrogen, the core will contract and the outer layers will expand – becoming a *red giant*. Stars of this size, having converted hydrogen to helium for billions of years, create a helium-rich core which becomes fuel for further building of heavier elements. Once the helium runs out, the core will expand and cool, turning into a white dwarf and eventually a black dwarf [6].

O-type stars are the largest stars with sizes up to 150 solar masses -- because of the high surface temperature they have a *blue color*. They have enormous pressures and temperatures in their core and thus burn their fuel rapidly – having lifetimes of *only a few million years*. They end their lives in a spectacular explosion – a supernova *(a major source of cosmic rays)* which becomes one of the key sources of *climate-change*.

Hydrogen protons fuse into helium in the core of the star. The big bang itself produced hydrogen and helium-4. The helium nuclei fuse together to produce carbon-12 and oxygen-16 nuclei. Each of these fusion reactions *turns* initial *energy of mass (m)* into further kinetic energy. [Energy of mass from Einstein's $E = mc^2$ where c is the speed of light.] The fusion of *iron* does not *produce further kinetic energy*, but *rather absorbs it*, and the fusion process goes no further in creating heavier elements [6]. (See Appendix B for a beginners look at protons, neutrons and electrons).

The nuclei in the core of the *massive star* become primarily iron and the core collapses. The core with such enormous mass falls into itself in a mighty implosion that produces a neutron star. The newly formed neutron star -- as described by Goldsmith [5] "produces a shock wave extending outward, reversing the inward fall of the star's outer parts and blasting them into space at speeds of thousands of miles per second. This supernova detritus includes not only nuclei lighter than iron, made before the explosion in relatively large amounts, but also heavier nuclei, fused in small quantities by the blast of the explosion itself."

The nuclear furnace of the stars provides the heavier elements. The remnants of those explosions contain the further important heavier elements, scattered about the universe as cosmic rays – and essential for human life[6]. The earliest stars are thus considered first generation stars and their remnants of atoms (often referred to as star dust) provided the basis for the creation of the second generation stars. Second generation stars comprise most of the stars in the universe today. These stars have 0.001 to 1% of their composition made up of elements heavier than those first produced in the big bang.

Our Sun is a third generation star with about 1-4% of the composition with the heavier elements. The solar wind (discussed in more depth later) is composed primarily of hydrogen (95%), helium (4%), and carbon, nitrogen, oxygen, neon, magnesium, silicon and iron (~1%).

The Milky Way galaxy is the home of our solar system. It is nearly as old as the universe itself (~ 13.6 billion versus 13.8 billion years). Neighbors to the Milky Way include the Large and Small Magellan Clouds, and the Andromeda galaxy; together with some 50 other smaller galaxies. This cluster is known as the Local Group.

Farther out is the Virgo Supercluster which includes the Local Group and another 100 galaxy groups – this has a 100 million light-year diameter. [One light-year is the distance traveled by light in one year: about 5.88 trillion miles.] This immense region is a source of potential cosmic rays interacting with Earth.

The Milky Way galaxy has a 'black hole' at the center of its disc as do most of the larger galaxies. A black hole has gravity within that is so strong (because matter has been squeezed to such a relatively tiny space) that light cannot escape. Our home galaxy has a diameter of 120,000 light years across and has a central bulge of 12,000 light years. It has considerable dust and gas (10-15% of the luminous/visible matter) with the rest being the stars themselves. In the visible spectrum one can only see 6,000 light years into the disk. However, infrared light can see through the dust and infrared telescopes have provided detail about the star birthing and decay within the Milky Way.

The solar system formed 4.6 billion years ago from the gravitational collapse of a spinning mass of hydrogen, helium and recycled star dust. As the spinning gas and dust flattened into a disc, aggregates circling the star formed into planets including Earth. Most of the hydrogen and helium has been used to form the Sun. The dust was very important for life on Earth as it was a mixture of iron, oxygen, hydrogen, carbon, nitrogen and many other elements essential for life[1].

Every effect has a cause, there has to be a first cause, many call that first cause God, creator of the universe and the big bang. Only a Supreme Being could have the wisdom to formulate all of creation -- from the subatomic particles in Appendix A to the massive galaxies with $10^{22}$ stars. It would take three generations of star formation to create the elements necessary for life on Earth.

Our solar system is located in a perfect location within the galaxy in terms of safety and visibility. The solar system is located within the galaxy, sufficiently far from the center of

the galaxy to be safe from *intensive* bombardment of cosmic rays from exploding stars, and yet conveniently placed between two star-filled spiral arms of the Milky Way galaxy-- away from the brightness of the massive stars and away from certain regions of thick dust clouds – either of which would hinder our excellent view of the cosmos.

The Milky Way galaxy has been placed in one of the darkest locations in the universe where intelligent life can exist. Thus, not only was the universe created for life on Earth, we have been purposely placed in a position to study and marvel at the magnificent universe created for us[8].

## References: Chapter 2

1.  Ross, H. N., 2018: *The creator and the cosmos.* 4th edition. RTB press, 335pp.

2. Tolman, R. C., 1922: Thermodynamic treatment of the possible formation of helium from hydrogen. *Journal of the American Chemical Society*, **44**, 1902-1908.
3. Gamow, G., 1946: Expanding universe and the origin of elements. *Physical Review*, **70**, 572-573.
4. Ade, P.A. R. and many others in alphabetical order, 2015: The Planck Collaboration. *Astronomy and Astrophysics.* **594**.
5. Guth, A. H., 1981: Inflationary universe: A possible to the horizon and flatness problems. *Physical Review.* **D 23**, 347-358.
6. Goldsmith, D., 2000: *The Runaway Universe.* Perseus Books, Cambridge, MA. 232 pp.
7. Ryden, B., 2003: *Introduction to cosmology.* Addison-Wesley.
8. Ross, H. N., 2008: *Why the universe is the way it is.* Baker press. Grand Rapids, MI. 240 pp.
9. http://www.sun.org/

# CHAPTER 3

# CO$_2$ and Climate Change in the Early Atmosphere

The Earth was formed 4.6 billion years ago. The process of chemical-density differentiation of Earth's matter led to a *gradual growth* of a dense iron-oxide core. This process occurred over the entire Archaean Eon (4 to 2.5 billion years ago). The young Earth had no hydrosphere. The sea basins were formed during the Early Archaean time[1].

The atmosphere was formed some 600 million years later. Nitrogen dominated the initial atmosphere, it had a partial pressure alone greater than one atmosphere at 4 billion years ago, a value of 1.4 atmospheres 2.5 billion years ago, and then slowly evolved to the current atmospheric value of approximately 78% by volume[2].

The Earth has been a warm wet volcanic planet over 80% of its history and only 20% have been designated as ice ages[1]. The peak of the CO$_2$ degassing rate coincided with the maximum tectonic activity about 2.7 billion years ago – reaching a value of approximately 10,000 times the current atmospheric value of 400 ppmv (parts per million by volume)[2].

Water degassing occurred much earlier, but reached its peak about 2.5 billion years ago, after the Earth had formed its high density iron-based core. The combination of the excessive CO$_2$ and available liquid water combined to capture CO$_2$ within carbonate rocks. The chemical reaction is provided below[2]:

$$2Ca \cdot Al_2 Si_2 O_8 + 4 H_2O + 2 CO_2 \rightarrow Al_4 [Si_4 O_{10}] (OH)_8 + 2 Ca CO_3 + 110.54 \, kcal / mole$$

$$\text{anorthite} \qquad\qquad\qquad\qquad \text{kaoline} \qquad\qquad \text{calcite}$$

The chemical equation is not really an equation, but a relationship. The law of stoichiometry requires that the same number of atoms of each element that appear on the *reactant side* (left side of the arrow) must appear on the *product* side (right side). [A brief introduction to chemistry is provided in Appendix B which will be more than sufficient for the purpose of this book]

A quick check of the above chemical reaction reveals: that two calcium (Ca) atoms are matched on both sides; four (2x2) aluminum (Al) atoms on the left are matched by 4 (Al) on the right; four (2x2) silicon (Si) atoms on the left are matched by 4 (Si) on the right; two carbon (C) atoms on the left match two on the right; and (24 = 16+4+4) oxygen (O) atoms on the left are matched by (24 = 10+8+6) (O) on the right.

There are other such formulae involving carbonates as above where for every two $CO_2$ molecules fixed in carbonates, four water molecules are used for hydration of the rock forming minerals in the oceanic and continental Earth crust[3]. The partial pressure of oxygen was quite low one billion years ago, as oxygen was consumed by the oxidation of iron. However, at ~500 million years ago after the disappearance of iron in the mantle, oxygen began to rapidly accumulate and provide the oxygen required for the highly organized life forms that developed at that time[2].

A comparison of $CO_2$ concentration with the 'well documented' ice ages over time is presented from data shown in Fig 3.1. This figure first appeared in the web site of "Plant Fossils of West Virginia" (www.geocraft.com/WVFossils/Carboniferous_climate.html) and appeared later in the popular book of Plimer[1]. The original reference for the carbon dioxide values in the figure are from Berner[3] and that of temperature from Scotese[4].

The black line in Fig 3.1 indicates the approximate history of $CO_2$. The time line of Fig 3.1 does not go back as far as desired, but there are other data sources in the scientific literature that do. This figure performs well over the past 500 million years. The blue line is an approximation of the temperatures at the time indicated. A very slight revision from Scotese has been made but not shown here.

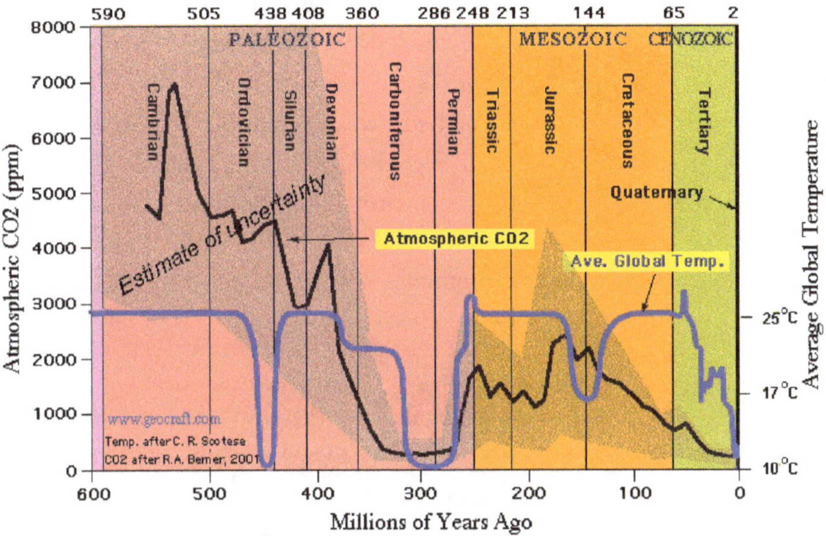


Figure 3.1  Estimates of the $CO_2$ concentration and the Earth's surface temperature over time

Atmospheric $CO_2$ continued to decrease (with oscillations up and down) until it had the value between 10 and 200 times today's concentration by 1.8 billion years ago[5]. Table 3.1 indicates $CO_2$ concentrations during the Earth's Ice Ages. The *oldest and coldest* well documented Ice Age was the Cryogenian Period (750 to 580 million years ago (MYA)). *The $CO_2$ concentration at that time was greater than 100 times the current value – 40,000ppmv.*

During the Karoo Ice Age (also called the Permo-Carboniferous) (350-280 MYA) the $CO_2$ concentration was *near the same value as today's value* (~ 370 to 400 ppmv). Other ice ages are shown in Table 3.1 where the values of carbon dioxide concentration lie between the two above extremes of 40,000 and 400 ppmv. *Clearly, $CO_2$ values have no correlation with the ice ages* – all the ice ages were due to external influences which are discussed in Chapter 12.

Table 3.1 also indicates the periods *between* the ice ages where the Earth's surface temperature was estimated to be ~ 6 degrees centigrade higher than today's average value. The Table indicates that there is *no correlation* of the $CO_2$ concentrations with these recurring warm periods! Those warmer periods occurred when the Sun's magnetic field was strong *or* the Earth's albedo was decreased. The Sun's insolation has virtually not changed over the past billion years – though it is ~ 40% higher than when it was formed 4.6 billion years ago[2].

Table 3.1  Ice Ages and Intermediate Warm Periods from 850 to<br>65 million years ago

| $CO_2$ (ppmv) | Various Ice Age Times (MYA)/name | Intermediate warm Periods (MYA) $\Delta T = {\sim}\,6\,°C$ | $CO_2$ (ppmv) |
|---|---|---|---|
| ~ 40,000 | 850 – 630 Cryogenic | 600 – 480 | 4200 |
| 4000 – 4400 | 460 – 430 Andrean-Saharan | 420 – 360 | 3000 |
| 370 – 400 | 350 – 280 Karoo | 240 – 170 | 1200 |
| 2000 – 2400 | 160 –120 Scutum-Crux | 100 – 65 | 1000 |

Table 3.1 also indicates the warm periods between the Ice Ages. The temperatures are estimated to have been approximately 6 °C warmer than today's temperatures, and the wildly varying $CO_2$ values range from 1000 to 4200 ppmv – *showing no correlation of $CO_2$ with the Earth's surface temperature during these interglacial periods.*

One can study data from ice cores over the past 420,000 years to obtain further *climate-change* information. Ice cores can provide oceanic water evaporation temperature from isotopic shifts in oxygen and deuterium, and $CO_2$ content in air bubbles to form time records of temperature. $CO_2$ Ice cores with sufficient vertical resolution (time resolution) have provided 420,000 years of data from Antarctica indicating that the *temperature changes preceded the corresponding $CO_2$ changes.* [More discussion of *light* oxygen and *heavy* oxygen can be found in Appendix B.]

An American team[6] went back 250,000 years and found the time lag (due to ocean mixing) of $CO_2$ behind temperature by 400 to 1000 years during all three glacial-interglacial transitions over that period. The oceanic reservoir of $CO_2$ is far greater than that of the atmosphere. When the oceans are warm, they outgas $CO_2$, and when the oceans are cold atmospheric $CO_2$ dissolves into the oceans via several different processes.

A subsequent study in 2003 by a French team indicating that deglaciation was not caused by $CO_2$ which lagged the temperature by 200 - 800 years[7]. A third effort by Russian scientists arrived at the same conclusion, where the estimated delay was 500-600 years[8]. This was claimed to be 420,000 years of data with *undisputable evidence* that $CO_2$ concentrations of the atmosphere are the *effect* of global temperature changes and not their *cause*[9].

## References: Chapter 3


1.  Plimer, I., 2009: *Heaven and Earth -- global warming: the missing science.* Quartet Books, Limited, London, 503 pp.
2.  Sorokhtin, O. G., G. V. Chilingar and L. F. Khilyuk, 2007: *Global warming and global cooling: evolution of climate on Earth.* Elsevier, Amsterdam, 313 pp.
3.  Berner, R. A. and Z. Kothavala, 2001: GEOCARB III: A revised model for atmospheric $CO_2$ over Phanerozoic time. *American Journal of Science,* **301,** 182-204.

4. Scotese, C. R., 2013: Paleomap project update. 2013 / Third Annual PaleoGIA & PaleoClimate Users Conference.

5. Kaufman, A. J. and S. Xiao, 2003: High $CO_2$ levels in the Proterozoic atmosphere estimated from analysis of individual microfossils. *Nature*, **425**,279-282.

6. Fisher, H., M. Wahlen, J. Smith, D. Mastroianni and B. Deck, 1999: Ice core records around the last three glacial terminations. *Science*, **283**, 1712-1714.

7. Caillon, et al., 2003: Timing of atmospheric $CO_2$ and Antarctic temperature changes across Termination III. *Science*, **299**, 1728-1731.

8. Monin, A. S., and D. M. Sonechkin, 2005: *Climate fluctuations based on observational data: the Triple Sun Cycle and other Cycles.* Nauka, Moscow. 191 pp.

9. Chilingar, G. V. et al., 2008: Greenhouse gases and greenhouse effect. *Environ Geol.* Springer–Verlag.

# CHAPTER 4

# Atmospheric Weather Variability

The atmosphere's tremendous weather variability suggests that a clear distinction be made between weather diversity and *climate-change*. It cannot be stressed enough that predicting the climate is a difficult scientific task. The largest scale instability of the atmosphere is itself chaotic – as will be shown. Weather is chaotic and not perfectly predictable. Many scientists in different fields are now aware of chaos since it was introduced and publicized by Edward Lorenz[1] in 1963.

Virtually every atmospheric and climate scientist is aware of chaos, however this book is also for the general public and an extremely short introduction to chaos may be useful. One cannot fully appreciate *climate-change* without that picture of chaos in one's mind. Two examples of chaos from equations that are *nonlinear quadratic* and *nonlinear cubic* will be shown – before demonstrating the impact of chaos in the large scale dynamics of the atmosphere.

A set of three nonlinear equations are sufficient to demonstrate chaos. The equations used by Lorenz demonstrate extremely simple convection of a two-dimensional fluid cell heated from below and cooled from above with the convection driven by three partial differential equations:

$$X(1)^* = P\,X(2) - X(1)$$
$$X(2)^* = -\,X(1)X(3) + RX(2) - X(2)$$
$$X(3)^* = X(1)X(2) - BX(3)$$

The equations are *nonlinear* and *quadratic* with two dependent variables multiplied together as in the 2nd and 3rd equations. The asterisk over the left hand side of the X-variables implies a time derivative (a rate of change over time). The definition of the X-variables from 1 to 3 and the constant terms P, R and B are not important here and can be examined in the reference if interested. Two views of the resulting trajectories from the numerical integration of these equations are shown in Figs 4.1 and 4.2.

Following the graphic drawing of these images in real time, one sees that the trajectories are bounded (they do not run off the page, nor do they repeat) – they produce a somewhat exotic double spiral in three dimensions. These have been refered to as strange attractors.

The concept of marrying the so-called Stochastic Dynamic Equations (SDE) with the Monte Carlo technique was performed in Fleming[2] where various examples of chaos from different dynamic sets of equations were examined in a deeper nonlinear perspective with the combining of the two analytical methods.

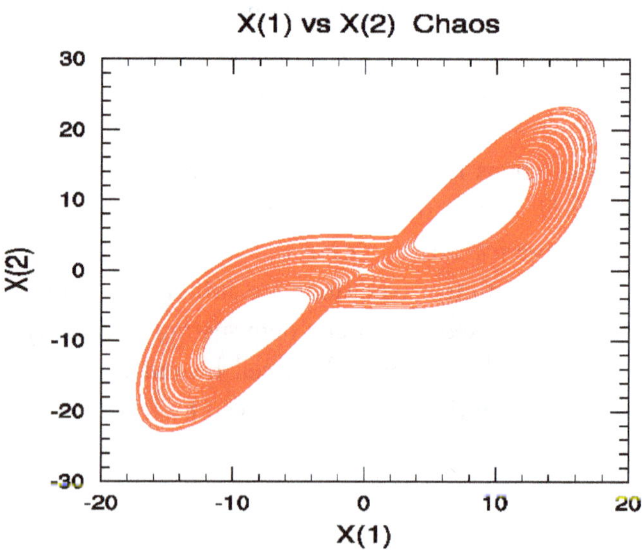


Figure 4.1  Lorenz trajectory X(1)-X(2) plane

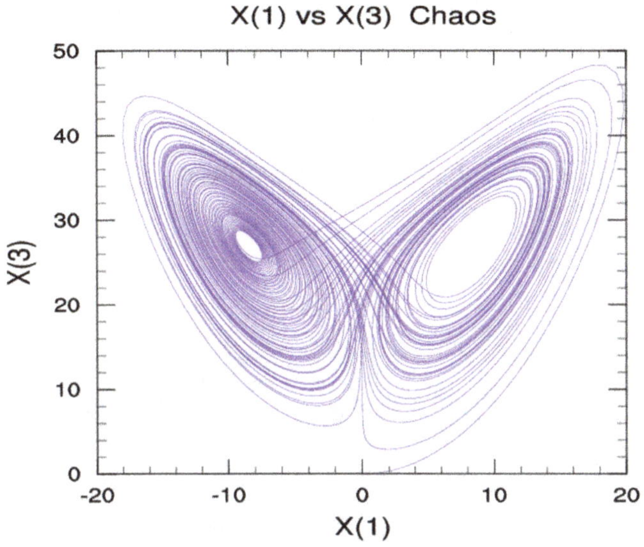


Figure 4.2  Lorenz trajectory X(1)-X(3) plane

The methods were of nonlinear quadratic and cubic systems. Proceeding to *nonlinear cubic systems* required the SDE equations be extended and they were derived in that reference.

The nonlinear cubic system was drawn from the work of Salazar and Nicolis[3] who used *two coupled oscillators* in an attempt to model ocean temperature / ice extent feedback. The author does not personally care for this type of modeling approach, would not recommend it, *and debated even mentioning it here*; but the equations offer a unique spectrum of exotic attractors — *just a few examples of many different types are shown here as window dressing for the reader.*

The set of nonlinear equations for these coupled oscillators are provided below.

$$X(1)^* = B1\ X(1) - A1\ X(2) + D1[X(3) - X(1)] - X(1)\ X(2)\ X(2)$$
$$X(2)^* = X(1) - X(2)$$
$$X(3)^* = B2\ X(3) - A2\ X(4) + D2[X(1) - X(3)] - X(3)\ X(4)\ X(4)$$
$$X(4)^* = X(3) - X(4)$$

Equations in lines 1 and 3 have the nonlinear cubic terms. *Changing the values of any of the constants produces a different attractor — see reference.*

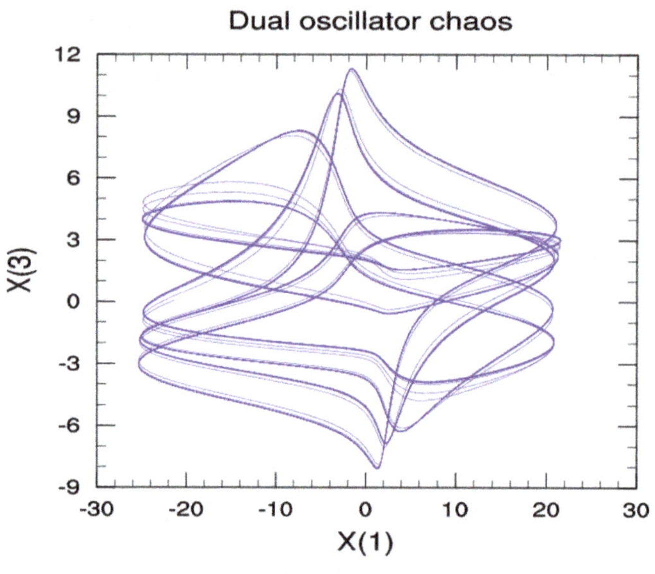


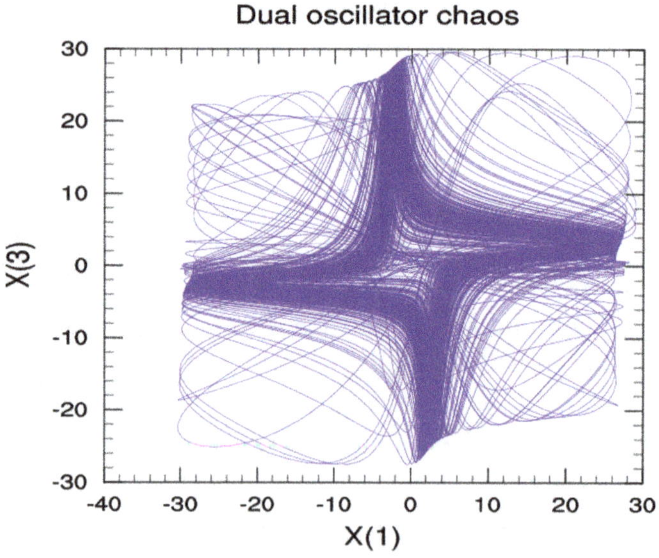

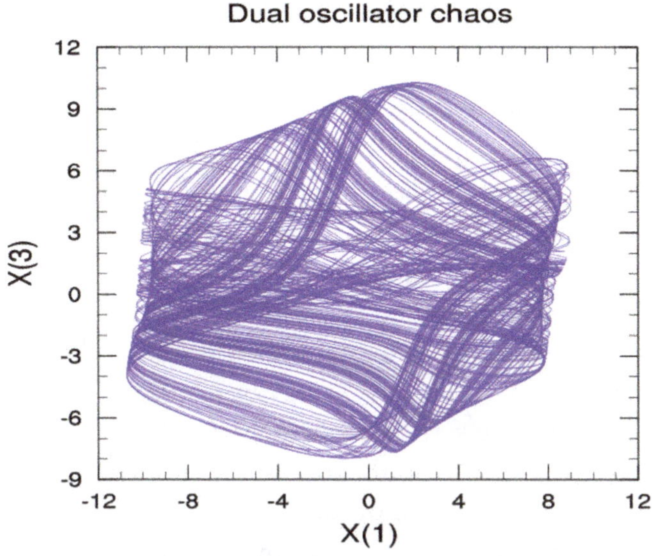


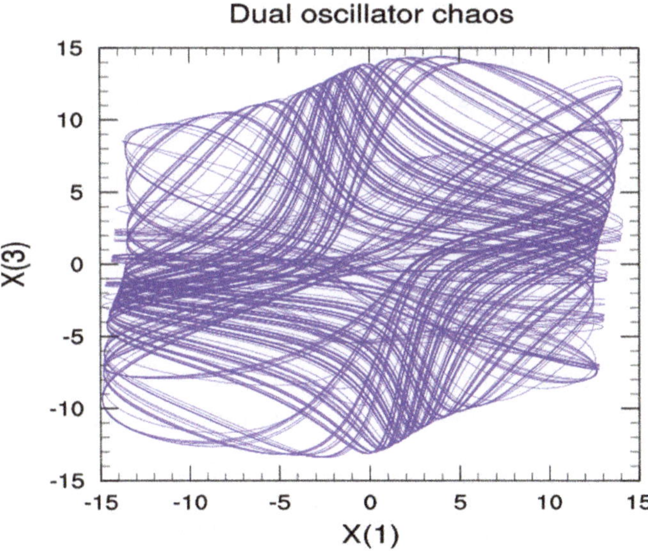


Figure 4.3  Various possible attractors produced from the coupled twin oscillator equations

Now let's get *back to serious business* and mention some details about the Earth's atmosphere – then move on to how large scale chaos contributes to tremendous weather variability on Earth. The atmosphere has a temperature profile decreasing with height through the troposphere which contains 80% of its mass. At the minimum temperature level of the troposphere is the tropopause – this varies with latitude and season. The tropopause height is generally 9 km over the poles and 16-17 km at the Equator.

Above the tropopause, the stratosphere temperature increases with height to about 50 km. The stratosphere obtains its heat from by the direct absorption of the Sun's energy by ozone ($O_3$). The stratosphere is cooled by longwave emission mainly by $CO_2$ and to a lesser degree by $H_2O$ and $O_3$. Above the stratosphere is the mesosphere extending from 50 to 85 km above the Earth. Temperature decreases with height and the coldest temperature in the atmosphere is -90°C at the top of this layer. Higher concentrations of iron and metal atoms occur in this layer due to the fact that meteors vaporize in this layer.

Most atmospheric gases have emission/absorption bands in the microwave region (well above the infrared region). These bands are not important for infrared radiation transfer in the atmosphere as the thermal radiative fluxes are very small. These bands are used to infer vertical profiles of temperature, moisture and liquid water in clouds through remote sensing from satellites[4].

The primary driver of Earth's weather systems is the very large scale process of *baroclinic instability* that occurs independently in both hemispheres. Differential heating between the incoming solar radiation and the outgoing infrared radiation creates a pole-to equator temperature gradient and produces a growing supply of available potential energy. Eventually, the *zonal thermal wind,* developing to balance that temperature gradient, becomes baroclinically unstable. The resulting large scale baroclinic waves transfer warm air poleward and cold air equatorward.

At the same time, the *eddy* (wave) available potential energy is converted into eddy kinetic energy by the vertical motion within the waves – maintaining the kinetic energy of the atmosphere against frictional dissipation. The waves intensify until the heat transferred poleward balances the radiation deficit. Various process within the atmosphere (friction, radiation to space) damp the unstable waves and the baroclinic cycle is repeated -- often referred to as a vacillation cycle.

Thompson[5] developed a low-order general circulation model consisting of a single finite amplitude baroclinic wave interacting with the zonal mean shear flow, maintained against a friction parameter (**D**) and driven by a differential heating term (**H**) – thus containing all the requirements for baroclinic instability. The model produced accurate values for certain features including the *vacillation period* of approximately 23 days – close to that seen in the Southern Hemisphere[6].

Vacillation, with fixed point attractors, were the only dynamic entities included and described in Thompson's original presentation. Later this model was found to produce two other attractor types: limit cycles and chaos (Fleming[7]). The chaos was produced over various values of the parameters **D** and **H**, and also produced by the process of *sensitivity to initial conditions* – whereby on a *strange attractor* two initially close trajectories on the attractor eventually diverge from one another – exponentially over time.

Fig 4.4 indicates two solutions in the above model, vacillation and chaos respectively, where X(1) represents the mean horizontal temperature gradient and X(2) represents the net poleward heat transport. The longest vacillation cycle was 25.2 days. The longest chaos cycle was 35.0 days. The trajectory of the solution forms a *strange attractor* on the top of Fig 4.4. This represents a kind of infinite complexity. The trajectory stays within a definite bounds, but never repeats itself.

The initial conditions were the same for all the parameters and variables – except that the initial mean horizontal temperature gradient [X(1)] was 67% of its fixed point value = 0.415 for the *vacillation* solution, and 66% of that value = 0.409 for the *chaos* solution. [Note that the theoretical values of the X(1) and X(2) *fixed points* (the final destination achieved in the center of the vacillation figure on the bottom), resulted from the formulas of the model parameters.][5]

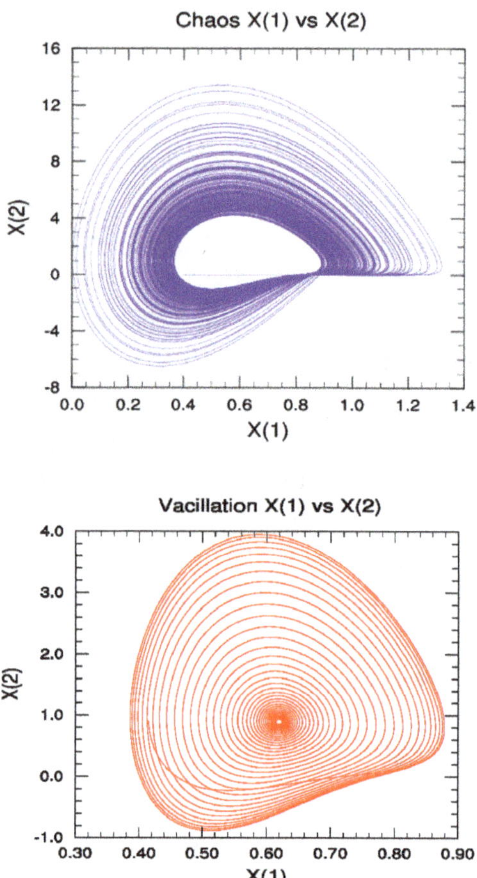

Figure 4.4  Explosive baroclinic Instability -- vacillation on the bottom and chaos on the top

Notice the significant difference in the scales of the two solutions in Fig 4.4. The important difference between vacillation and chaos is that the range of X(1) is approximately three times greater in the X(1) direction and five times greater in the X(2) direction for the chaos case – the complete trajectory of the vacillation solution nearly fits inside the opening in the chaos trajectory. When such a large solution difference occurs from such two closely spaced *initial points*, this is *Explosive Baroclinic Instability (EBI)*.

*A Monte Carlo approach evaluated the power of the chaos within this nonlinear model of baroc*linic instability. Using a known chaos initial state for X(1), 40,000 different initial states were selected from a random number generator for a normal distribution with a standard deviation of only 0.001. The only model value changed for the 40,000 different deterministic runs was *X(1) = 0.4 + the normal deviate.*

The results for the 40,000 deterministic chaotic solutions were all different. The spread of solutions, using X(3) as an example (the wave kinetic energy -- considered a proxy for storm intensity), provided a dynamic range. The maximum value of X(3) within a chaos run was considered as a measure of the strength of that run. The *average* maximum X(3) for all the 40,000 runs was 18.95. The minimum and maximum of this X(3) measure were 7.42 and 27.68 – nearly a factor of four difference in magnitude.

This weather diversity would expand with the seasons due to different heating characteristics. This model underestimates the degree of variability. If the model contained the 5-7 waves typically seen in a hemisphere, the nonlinear wave-wave interaction would have created greater weather variability. This weather diversity would occur within any *climate-change* regime – warm or cold. However, there will never be *runaway chaos* as EBI is limited by the dynamics of the system[8]. Nevertheless, weather variability is such that there is a high probability that some weather parameter may experiences a record somewhere on the planet on any given day. The media often get carried away by such records, inferring climate change – not so.

Further changes in **H** and **D** were evaluated over a wide range. Large **H** and small **D** provide a fast system and small **H** and large **D** a sluggish system as anticipated. However, smaller values of **H and D**, as might be expected within an ice age or colder periods, lean toward more chaos[2].

## References: Chapter 4


1. Lorenz, E. N., 1963: Deterministic nonperiodic flow. *J Atmos. Sci.* **20**, 130-141.

2. Fleming, R. J., 2015: *Analysis of the SDE/Monte Carlo approach in studying nonlinear systems.* Lambert Academic Publishing, Saarbrücken, Germany, 136 pp.

3. Salazar, J. M. and C. Nicolis, 1988: Self-generating aperiodic behavior in a simple climate model. *Climate Dynamics*, **2**. 105-114.

4. Peixoto, J. P. and A. H. Oort, 1992: *Physics of climate.* American institute of Physics, New York, 522 pp.

5. Thompson, P. D., 1987: Large-scale dynamic response to differential heating: statistical equilibrium states and amplitude vacillation. *J. Atmos. Sci.*, **44**, 237-1248.

6. Webster, P. and J. L. Keller, 1975: Atmospheric variations: vacillation and index cycles. *J. Atmos. Sci.*, **44**, 1238-1300.

7. Fleming, R. J., 2014: Explosive baroclinic instability. J. *Atmos. Sci.*, **71**, 2155-2168.

8. Finding the roots of the characteristic equation of a nonlinear system, provides the eigenvalues. The sum of the real parts of these roots is the "trace". If the trace is negative (as is the case for the model in the text which produces explosive baroclinic instability (EBI), there will never be runaway EBI -- the chaos is constrained.

# CHAPTER 5

# CO$_2$ and Climate Change: 250 BC through to 1850 AD

Ian Plimer is a famous geologist from Australia and makes some very important points in his latest book[1] about how most people do not look upon *climate-change* from a proper perspective. He quotes "Geology is about time, changes to our environment over time and the evolution of our planet. Geology is the only way to integrate all aspects of our environment". He points out "that there has been little or no geological, archeological and historical input into discussions about *climate-change*".

It is necessary to respect his views and this section will draw upon some of the historical facts that he has reproduced. During the last 730,000 years there have been major glaciation periods – separated by interglacial periods. Records of these glaciations are seen in deep sea sediment cores in all the oceans. The last glaciation started 116,000 years ago an ended 14,000 years ago. Sea level then rose 130 meters at a rate of one cm per year[1]. The alternating temperature changes were global in nature.

One must also consider the climate change over the past 14,000 years that the interglacial period has been in place for Earth – a reprieve from our Current Ice Age which began approximately 2.7 million years ago. He has described many different *climate-changes* that go from cold to warm to cold again over the time period of 11,600 before present to the end of the Little Ice Age in 1850. Here we shorten the list to an abbreviated Table from his book and just discuss a few of these warm /cold periods.

Table 5.1  List of significant periods of warm and cold
(bp = years before present)

| Climate Changes | Dates of changes |
| --- | --- |
| Holocene Warming a | 11,600 – 8,500 bp |
| Egyptian Cooling | 8,500 – 8,000 bp |
| Holocene Warming b | 8,000 – 5,600 bp |
| Akkadian Cooling | 5,600 – 3,500 bp |
| Minoan Warming | 3,500 – 3,200 bp |
| Bronze Age Cooling | 3,200 – 2,500 bp |
| Roman Warming | 500BC – 535AD |
| Dark Ages | 535AD – 900AD |
| Medieval Warming | 900AD – 1300AD |
| Little Ice Age | 1300AD – 1850AD |
| Modern Warming | 1850AD –......... |

Many summaries of alternating warm and cold periods occurred within that interglacial period and hundreds of references exist on the details. The last glaciation ended 14,000 years ago, followed by a post- glacial warming lasting 1500 years. This was followed by the intense cold period from 12,900 to 11,600 -- the Younger Dryas. Just a brief review of further warm and cold periods are provided here with some of the data sources[1].

Pollen grains (seeds from trees and flowering plants) are preserved in sediment layers of lakes. Lake sediments in Peru provided a 4000 year record indicating the *Roman Warming* (500 BC to 535 AD), less pollen in the *Dark Ages* (535 to 900 AD), increased pollen in the

*Medieval Warming* (900 to 1300) and a pollen decline in the *Little Ice Age*. Cave stalagmites with carbon and oxygen isotopes provide temperature information. Tree rings provide a temperature record – a 1300 year record shows the warmest decades in the Medieval Warming and the coldest during the Little Ice Age. Other records can come from ship logs and written accounts of military campaigns.

*Roman Warming* had grapes grown in Rome in 150 BC. By the first century BC Roman historians reported vineyards and olive trees extending northward within Italy[2]. Europe enjoyed a Mediterranean climate. By 300 AD the global climate was far warmer than at present[3]. Central America and Central Asia had warmer, wetter weather and a strong population increase.

Researchers drilling in the crater lake of an extinct volcano, Mt Kenya, recorded data from 2250 BC to 750 AD and found that the period 350 BC to 450 AD was a significant high temperature period. This was the *Roman Warming* reflected as a warmer climate in equatorial Africa[4].

The *Dark Ages* saw a sudden cooling begin in 535 and 536 which continued until 900[5]. The Black Sea froze and ice formed on the Nile River – this level of freezing has not happened in that area since then. The lack of sunlight and drought caused crop failure and famine followed[6].

*Medieval Warming* (900 to 1300) was a warm period where society thrived! The summers were long, the crops were plentiful and there was food for all. The population increased, cities grew, universities were established and cathedrals were built[7]. A study of 6000 boreholes (as rock transmits past temperatures downward for ~1000 years) obtained from all the continents indicated that the *Medieval Warming* was warmer than today[8]

The *Little Ice Age* (1300 - 1850) occurred in two major phases, with the second phase being the coldest of any period since the last glacial period[9]. Famine in Europe killed millions between 1690 and 1700, further famines occurred in 1725 and in 1816[1]. The cold climate was global.

What made the Little Ice Age much more difficult were the years of warmth in the Medieval Warming where there was a tremendous population increase[1]. The population had naturally adapted to warm times and were not prepared for the sudden onset of cold times. This apparently contributed to the massive depopulation.

The *Little Ice Age* was initiated with what is called a "quiet Sun" a period of very few sunspots – referred to as a *Solar Minimum.* There was a sequence of Minima during the *Little Ice Age* – the most important were the Maunder (1645 – 1715), and the Dalton (1795 – 1825)[1].

Cosmic rays are star dust – mostly hydrogen protons from exploding stars. *These enter the Earth's atmosphere when the solar wind and its magnetic field are weak – when the Sun is "quiet".* They will be discussed further in Chapter 12.

An important radioactive isotope is produced from the normal light element of beryllium (with 4 protons and 5 neutrons) into beryllium-10 – it is produced by cosmic rays as follows. A cosmic ray entering the atmosphere creates a shower of secondary cosmic rays, e.g. an energetic neutron. This collides with an oxygen atom, removing a neutron to be added to beryllium to make beryllium-10 (4 protons and 6 neutrons). Be-10 has a half-life of 1.4 million years.

Fig 5.1 indicates Be-10 as a good proxy for solar magnetic field strength (strong magnetic field, more sunspot activity versus low Be-10 values). The Be-10 values are from Beer, et al.[10] and the sunspot numbers from Hoty and Schatten[11]. Be-10 values are high when cosmic rays are strong (when the Sun is weak) -- note the strong values during the deep and important Maunder Minimum of the *Little Ice Age* (1645 – 1715). Be-10 values have weakened considerably with the strong magnetic field and increased sunspot activity Sun during the *Modern Warming.*

Another proxy for solar activity involves cosmic ray interaction with nitrogen (N-14, normally 7 protons and 7 neutrons). In this case the energetic neutron collides with nitrogen and the atom *loses a proton and gains a neutron* to become Carbon-14 (6 protons and 8 neutrons). Normally the carbon atom is Carbon-12 (6 protons and 6 neutrons).

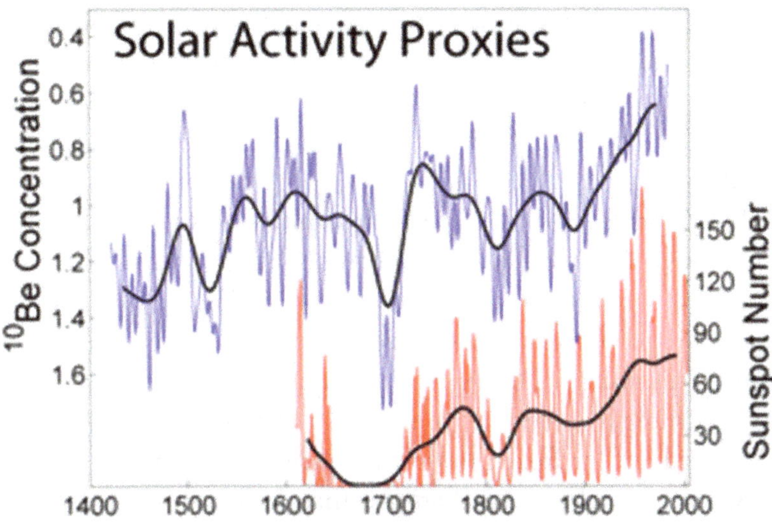


Figure 5.1  Be-10 values from Beer[10], et al. – sunspots from Hoyt & Schatten[11]

Fig 5.2 is adapted from two sources[12] and indicates *C-14* data are extremely small during the warm *Medieval Warming* and also clearly illustrate the Modern Warming which began after the end of the *Little Ice Age*. The C-14 values also have their largest magnitudes at the sunspot minima (when cosmic rays are abundant). Those maximum values match the phases of the *Little Ice Age*.

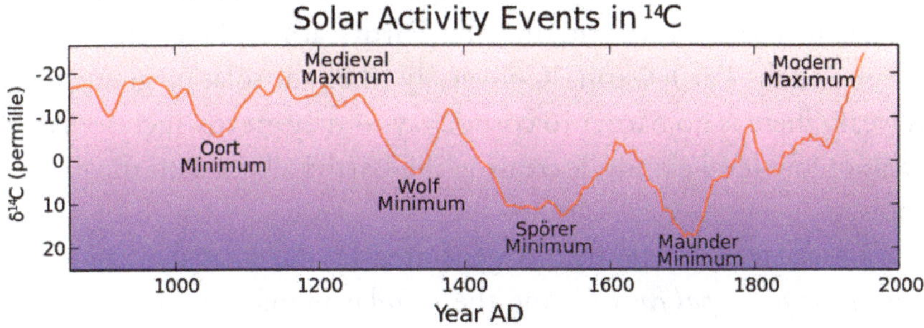


Figure 5.2  C-14 values (strong/weak) when cosmic rays are (strong/weak)

Oxygen isotopes of heavy and light oxygen in cave stalagmites in Ireland[13] confirmed the timing of the Roman warming, the Dark Age cooling and the Medieval Warm period where Greenland was + 6° warmer than today. [Information about *light* and *heavy* oxygen is in Appendix B.]

The *Little Ice Age* occurred in two phases with four quite cold periods at each of the four sunspot minimum periods – which were mentioned above (the Dalton is not shown in Fig 5.2). All four cold periods were confirmed by oxygen isotopes from those same cave stalagmites in Ireland[13]. During a brief period of warmth around 1500, ships returning to Greenland found that the entire Viking population had starved and/or frozen to death[1]. The second phase of the Little Ice Age was much colder than the first, especially in the 70 year Maunder minimum centered in 1680.

Since the *Little Ice Age*, a strong Sun is revealed by both Be-10 and C-14 decreases. The total magnetic flux leaving the sun dragged out by the solar wind) has risen by a factor of 2.3 since 1901 (Lockwood et al.)[14] – published in 1999. The strong solar magnetic field has shielded the Earth from cosmic rays and is the cause of the Modern Warming that has continued through to the current time. This competing climate theory is from Svensmark and Friis-Christensen[15].

*The new theory involves the interaction of the solar magnetic field with cosmic rays.* When the solar magnetic field is strong, it acts as a barrier to cosmic rays entering the Earth's atmosphere, clouds decrease and the Earth warms. Conversely when the solar magnetic field is weak, there is no barrier to cosmic rays – they greatly increase large areas of low-level clouds, increasing the Earth's albedo and the planet cools.

The solar magnetic field is generated by the solar dynamo with *one of the principal causes* being the angular momentum of the Sun's *differential rotation* (Charbonneau[16]). The Sun's equatorial region rotates faster -- 24 days, compared to the polar-regions which rotate once in ~ 30 days. This solar dynamo accounts for the variability of the sunspot amplitudes and frequency changes.

Another factor affects the solar dynamo that occurs on a longer time scale. This is the Sun's motion about the center of mass of the solar system – the solar system barycenter *(SSB)*. The position of the SSB is constantly changing primarily as a function of the mass of the Sun and the mass of the four major planets. Many more details are provided by Sharp[17] and discussed later in Chapter 12. The Sun's travel about the SSB adds its *orbital angular momentum* from that journey to its own *rotational angular momentum* so that *both contribute to important changes in the Sun's magnetic field intensity.* In keeping with our travel guide routine, these details did not surface until Svensmark's theory in 1997, the result of Lockwood in 1999, and other details in the last 50 years will be openly revealed in Chapter 12.

This new climate-change theory competes with the $CO_2$ warming associated with the timing of the Industrial Revolution. The combination of both of these two solar magnetic field influences appear to be the cause of a *20th century cooling within* the Modern Warming. Thus, it is imperative to also consider the 20th century temperature record *since* the Industrial Revolution.

Before progressing further with climate change observed within the 20th century, it will be informative to see the history of the $CO_2$ warming theory as it had evolved around the 1900 time frame – and the contributions of three important scientists at that pivotal point in time.


## References: Chapter 5


1. Plimer, I., 2009: *Heaven and Earth -- global warming: the missing science*. Quartet Books, Limited, London, 503 pp.
2. Allan, H. W., 1961: *The history of wine*. Faber & Faber, London.
3. Lamb, H.H., 1977: *Climate, history and the future*. Methuen, London.

4. Rietti-Shati, M., A. Shemesh, and W. Karlen, 1998: A 3,000 year climate record from biogenic silica oxygen isotopes in equatorial high-latitude lake. *Science,* **281**, 980 – 982.

5. Keys, D., 1999: Catastrophe: an investigation into the origins of the modern world. Ballantine.

6. Mango, C., 1980*: Byzantium, the empire of new Rome.* Scribner, New York.

7. Gimpel, J., 1961: *The cathedral builders.* Grove Press, New York.

8. Huang, S., H. N. Pollack and P.Y. Chen, 1997: Late Quaternary seen is worldwide continental  heat flow measurements. *Geophysical Research Letters,* **24**, 1947 – 1950.

9. Grove, J. M., 1988: *The little ice age.* Cambridge University Press.

10. Beer, J., St. Baumgartner, et al., 1994: Solar variability traced by cosmogenic isotopes in *The Sun as a variable star: Solar and Stellar Irradiance Variances* (eds. J. M. Pap, C. Fröhlich, H. S. Hudson and S.K. Solanki). Cambridge University Press, 291-300.

11. Hoyt, D. V. and K. H. Schatten, 1998: Group sunspot numbers: A new solar activity reconstruction. Part1. *Solar Physic,* **179**, 189-219.

12. Figure created by Lelind McInnes, Licensed under GNU Free Documentation License; also found in Plimer in a different black and white version.

13. McDermott, et al., 2001: Centennial scale Holocene climate variability revealed by high-resolution speleothem $\partial^{18}O$ record from SW Ireland. *Science,* **294**, 1328-1321.

14. Lockwood, R., R. Stamper and M. N. Wild, 1999: A doubling of the Sun's coronal magnetic field during the past 100 years. *Nature,* **399**, 437-439.

15. Svensmark, H. and E. Friis-Christensen, 1997: Variation of cosmic ray flux and global cloud coverage – a missing link in solar-climate relationships. *Journal of Atmospheric and Solar-Terrestrial Physics,* **59**, *1225-1232.*

16. Charbonneau, P., 2014: Solar dynamo theory. *Annu. Rev. Astron. Astrophys.,* **52**, 251-290.

17. Sharp, G. J., 2008: Are Uranus and Neptune responsible for Solar Grand Minima and solar cycle modulation? First published on line at http//www.landscheidt.info. arxiv.org/ftp/arxiv/Papers/1005/1005.5303.pdf.

# CHAPTER 6

# Arrhenius, Angström and Planck

All three of the above gentleman were considered excellent scientists working in their respective fields in the general 1900 time period. Their accomplishments relative to the subject of this book will be revealed – as well as their errors, though these errors were only due to the lack of proper data that was available to them at that time. These omissions will also be summarized.

Svante Arrhenius was a scientist born in Sweden in 1859 and who died in 1927. He was awarded the Nobel Prize in Chemistry in 1903. In 1896, Arrhenius published a paper[1] suggesting that a doubling of $CO_2$ in atmosphere would lead to an increase of 5 to 6 ° C. He had relied on earlier work of previous scientists and on available absorption coefficients from Knut Angström.

Arrhenius later wrote in 2008[2] that with the increases of $CO_2$ would provide some good with the bad: "… we may hope to enjoy ages with more equable and better climates, especially as regards the colder regions of the earth, ages where the earth will bring forth much more abundant crops than at present, for the benefit of rapidly propagating mankind." His calculations suggested that it would take 3000 years for industrial emissions at 1896 levels to double $CO_2$, and that this would occur because the oceans would absorb most of the extra $CO_2$

This theory of $CO_2$ climate warming caught on briefly, but then came to a quick end in 1900 when Knut Angström, son of the more famous scientist Prof. Anders Angström, published his 1900 paper[3] indicating that his measurements suggested that adding more $CO_2$ would not add more heat -- as the wavelengths of infrared absorption for $CO_2$ were already saturated and strong enough to absorb 100% of the radiation at those wavelengths. He further pointed out that the wavelengths for $CO_2$ overlapped rather significantly with those wavelengths of $H_2O$ which absorbed longwave radiation. This pretty much ended the discussion of $CO_2$ warming over the next 70 + years.

Arrhenius was correct about the value of increased warming to those colder countries of the world in his time (the warming would come from solar causes – shown in Chapter 12, not from $CO_2$). He was also extremely astute to realize the value of increased carbon dioxide for world food production for the ever increasing world population. He was wrong about requiring 3000 years for $CO_2$ to double – over estimating the ocean's ability to absorb the extra carbon dioxide, and under estimating the growth of $CO_2$ by the extent of the Industrial Revolution.

Angström was correct about the saturated absorption lines that he measured, but did not have access to all the lines we have today. He was unaware of all the lines available for $CO_2$ and $H_2O$ and overestimated the amount of overlap. However, neither the saturation issue nor the overlap aspect of the lines are important – the real concern is the radiation intensity. The radiation intensities of $CO_2$, $H_2O$ and all radiating gases, fall off dramatically with height (temperature) as indicated in Chapter 11. This important point was brought home by the work of the third important scientist of this period, Max Planck -- the greatest contributor to the study of radiation.

Max Planck was born in Germany in 1858 and died in 1947. His earliest work was on the subject of thermodynamics. He published papers on entropy, thermoelectricity and the theory of dilute
solutions. He was interested in radiation processes and was led to the problem of the distribution of energy in the full spectrum of the radiation from the Sun.

Planck was able to derive the relationship between the *energy intensity* and the frequency (or wavelength) of radiation. Planck provided the relationship in his paper published in 1900[4] which was based upon the revolutionary idea that the energy emitted by a resonator could only take on *discrete values or quanta.* This was Planck's most famous work, but more important it marked the turning point in the history of physics.

Though not considered the "father" of quantum mechanics, this was a first step towards a new quantum theory which has since solved many problems that Newton's classical physics could not handle.

The most important point for the purpose of this book is that the Planck function (formula) which will be used extensively in Chapter 11 is a very strong function of both *radiation wavelength ($\lambda$)* and *temperature.* The Plank function expresses the intensity of radiation! Had this function been firmly established in the heads of Arrhenius and Angström there would have been far more agreement between the two. This Planck function will be the key to showing why $CO_2$ has no role in climate change as demonstrated in Chapter 11. All three of these scientists also lacked today's knowledge of the Sun's powerful magnetic field, one of the keys to the cause of *climate-change.*

The view that $CO_2$ was not a cause of climate change was confirmed in the Compendium of Meteorology in 1951 of the American Meteorological Society by C. E. P. Brooks (1888 -1957).

The Earth's orbit can influence climate as addressed by Milankovitch[5] as will be shown below. There are three variations in the Earth's orbit with periods of 100,000, 41,000 and 21,000 years. These Milankovitch cycles were the consensus of what was believed to be the cause of *climate-change* in the 1970s. None of these three can routinely and *independently* explain Earth' dominant Ice Ages[6]. However, further progress on solar influences may yet couple with combinations of these periods to produce transitions between ice age periods and interglacial periods.

The Earth's eccentricity is the shape of the Earth's orbit around the Sun – it is not circular. The path varies from lightly elliptical

(0.005) to slightly more so (0.058) over a cycle of 100,000 years[6]. This altered the distance of the Earth from the Sun providing a 6% change in solar activity between January and July.

The Earth's axis is tilted, the inclination of the axis in relation to its plane of orbit about the Sun. The tilt varies from 24.5 to 21.5° on a period of 41,000 years. This *obliquity* changes the solar insolation energy difference received between the equator and the poles.

The Earth wobbles as it spins on its axis like a spinning top. This slow wobble is due to the gravitational pull of nearby planets and the moon. This is referred to as *precession* and has a periodicity of 21,000 years – it exaggerates seasonal contrasts[6].

Previous work using the shared variance at the *obliquity* and *precession* frequencies led to a limited correlation with Northern Hemisphere insolation and ice *volume (V)*; e.g., Hays[7]. Further progress was made by Edvardsson[8], et al. using the solar radiation power at solar solstice and comparing this with the *time rate of change of ice volume* (dV/dt).

Perhaps the best work was performed even later in time -- an extremely good correlation was obtained by Roe[9] by using regression analysis to obtain a best-fit linear combination of *obliquity* and *precession* parameters for the *change of ice volume with time (dV /dt)*. This excellent correlation is shown in Fig. 6.1

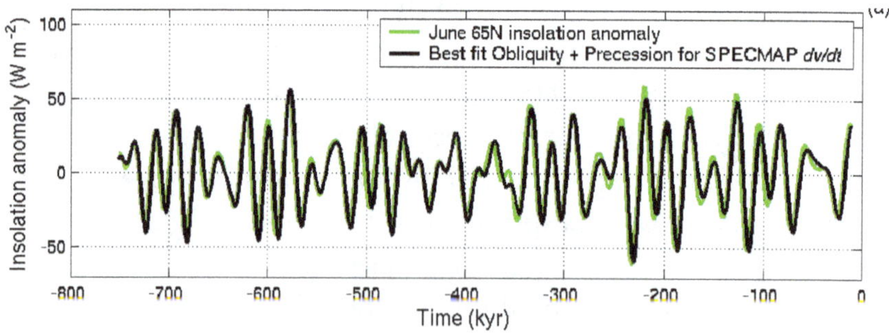


Fig. 6.1  Change in ice volume with time versus solar insolation in Northern Hemisphere

Angström's view remained unchallenged for nearly 80 years—then in the late 1970's Arrhenius' theory surfaced again – perhaps stimulated by a European government official. [The documentary film[10] "The Great Global Warming Swindle" alleged that the UK Prime Minister initiated research funds to prove that $CO_2$ (fossil fuel burn) was a problem – due to her fight with the coal labor unions and a desire for more nuclear energy.]

Many scientists in the UK, and eventually other scientists in other countries reacting to similar government actions, jumped at these new funds, competing for research grants and computer upgrades – even some scientists that had specifically warned in the early 1970's *that a new ice age was approaching*. [This was the beginning of a cool period within the Modern Warming discussed in the following Chapter.]

## References: Chapter 6


1. Arrhenius, S. A., 1896: On the influence of Carbonic Acid in the air on the temperature of the ground. *Plil. Mag.* Ser 5, **41**, 237-276.

2. Arrhenius, S. A., 2008: *Worlds in the making.* Harper & Row, New York.

3. Angström, K. J., 1900: *Intensity of solar radiation at different altitudes.* Nova Acta, 34 pp.

4. Planck, M., 1900: On an improvement in Wein's equation for the spectrum, Translated by Dirk ter Haar in *The Old Quantum Theory*, Pergamin Press, 1967.

5. Milankovitch, M., 1941: Canon of insolation and the ice age problem. *Akad. Beogr. Spec. Publ.* **132**, (translated by the Israel Program for Scientific Translations, Jerusalem, 1969).

6. Plimer, I., 2009: *Heaven and Earth -- global warming: the missing science.* Quartet Books, Limited, London, 503 pp.

7. Hays, J. J. Imbrie and N. Shackleton, 1976: Variations in the earth's orbit: pacemaker of the ice ages. *Science*, **194**, 1121-1131.

8. Edvardsson, S., et al., 2002: Accurate spin axes and solar system dynamics: Climate variations for the Earth and Mars. *Astronomy & Astrophysics*, **384**, 689-701.

9. Roe, G., 2006: In defense of Milankovitch. *Geophysical Research Letters*, **23**.

10. Do internet Google search for The Great Global Warming Swindle – Top Documentary Films.

# CHAPTER 7

# Modern Warming and the 35-year Cool Period Within

This new *climate-change* theory competes with the $CO_2$ warming associated with the timing of the Industrial Revolution. Fortunately, there is a distinguishing feature of a 35-year cooling event within the Modern Warming. There are two solar signals that compete with the $CO_2$ signal during this period. Clearly, it is imperative to consider the 20th century temperature record since the Industrial Revolution in detail.

The $CO_2$ concentrations have relatively little change over the interglacial period (see Fig 3.1). A Greenland ice core provided the value 270 ppmv dated from 600 years ago (Neftel et al. 1983)[1]. The pre-Industrial Revolution estimate by several methods provides a narrow range of 270 to 290. Thus there is no correlation of $CO_2$ with climate-change in the 11,500 interglacial period.

The World War II and post war period was a time of tremendous industrial growth from 1940 to 1975[2]. Fig 7.1 indicates the modern $CO_2$ record of the Mauna Loa Laboratory near Hilo, Hawaii which indicates the rise in $CO_2$ as a result of the 330% rise in hydrocarbon use. The benchmark record of carbon dioxide growth is the $CO_2$ record obtained from the Mauna Loa Laboratory. This record clearly showed the rise in the $CO_2$ concentration from the tremendous industrial growth of that period. The history of this Mauna Loa $CO_2$ data record is also informative.

47

Oceanographer Roger Revelle had performed research in the early 1950's and did not believe the ocean uptake of $CO_2$ would be anything like that stated by Arrhenius. Revelle then became more interested in this possible warming by $CO_2$ and hired Charles Keeling to start measuring atmospheric $CO_2$ on a high mountain laboratory near Hilo, Hawaii. The annual cycle evident in Fig 7.1 is the resulting Keeling curve indicating the rise of $CO_2$ since 1958[3].

One can continue the Keeling curve back to 1940 with data from Callendar[4] which showed the annual $CO_2$ concentration increase from 295 ppmv in 1900 to 310 ppmv in 1940. This gradual increase of 15 ppmv over the 40 year period roughly matches the general slope of the Keeling curve. The seesaw effect in the data reflects the changes of seasons within the annual cycle.

There was a significant drop in temperature from coastal stations around the Arctic Ocean from 1940 to 1970 of 1.4 °C. This was reported in a figure attributed to Dr. Akasofu, Director of the International Arctic Research Center of the U. of Alaska[5]. (The figure had the reference to the data as Polyakov et al. 2002, but the date may be slightly incorrect). Later in Chapter 8 there is another reference to the cool period from 1940 through 1980 in this same polar region.

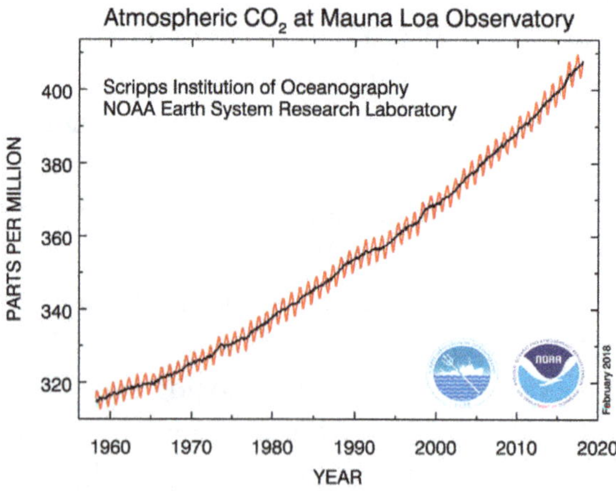


Figure 7.1  $CO_2$ record from Mauna Loa in Hawaii

Fig 7.2 also shows a *global cooling* between 1940 and 1975 of surface temperatures over land (90N to 60S) from three different records. These records are considered *quite accurate* by scientists on both sides of the *climate-change* issue – caused by $CO_2$ or not caused by $CO_2$. All three records also indicate the warming periods on either side of the cool event within the Modern Warming.

These temperature records of these three data sources have been shown many times in the scientific literature published in *support of the $CO_2$ climate warming theory.* However, the discussion of the cool period within the record is seldom, if ever, mentioned in that literature.

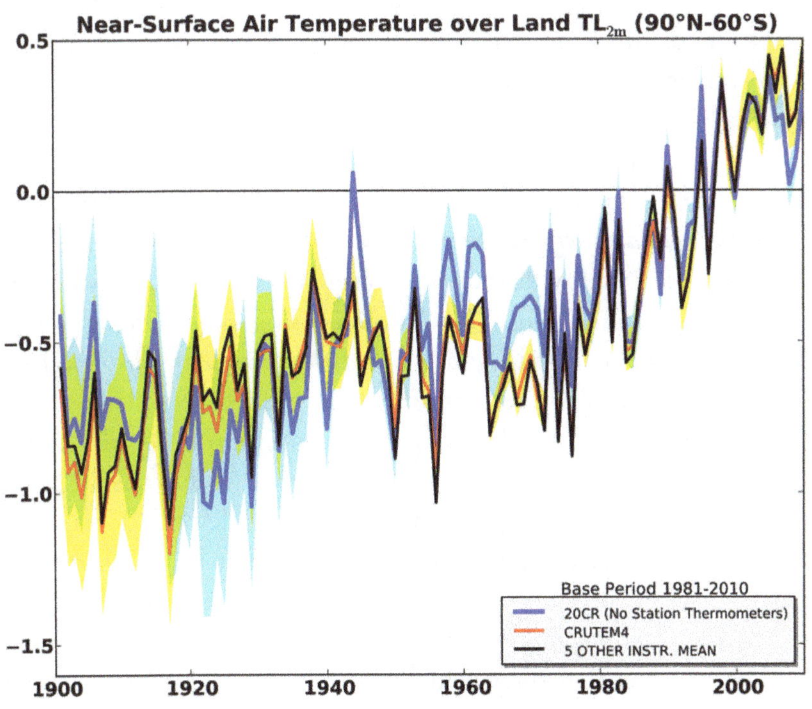


Figure 7.2 Near-surface air temperature change in the 20th century

The benchmark indicator for the $CO_2$ warming theory, the NOAA Mauna Loa carbon dioxide record, fails to indicate this 35-year period of cooling. There are two solar related events that do support this cool *climate-change* event.

In contrast to the *smoothed decrease* in the Be-10 data shown by the solid line in Fig 5.1 from the post Little Ice Age — *a closer look at the background shows sharp spikes of significant Be-10 increases from 1900 to 1970* indicating increased cosmic rays. A second solar record below provides further evidence of this cool period.

The Sun's sunspot cycle (the solar dynamo is discussed in Chapter 12) has an average period of 11.2 years, but the length varies from 8 to 14 years. The length of a sunspot cycle (LSC) is an indicator of the Sun's eruptional activity. The Gleissberg cycle resulted from the smoothing of this time series of the *length of the sunspot cycles (LSC)* and a secular cycle of 80-90 years emerged (this 85 year cycle is seen from 1890 to 1975.

Fig 7.3 is from (Landscheidt[7]) where Gleissberg's smoothed data was displayed. The heavy line is the smoothed LSC line and the light line is the *land air temperature in the Northern Hemisphere.* The heavy line agrees very well with the temperature decline and also with the temperature record of Fig 7.2 with the cooling from 1940 to 1975. The Gleissberg record had gone back 300 years using historical recorded aurorae data before sunspots were routinely recorded.

We have seen no correlation of $CO_2$ concentrations with any *climate-change* -- including the 35 year cool period within the 20th century. How can this be? Is the financial/political side of $CO_2$ warming propaganda so overwhelming that science turns it head and ignores the lessons of history? Let's examine a few facts with an open mind to see why the debate could even occur.

First, the fact that the world began to warm up about the same time as the Industrial Revolution *could* lead one to believe that increased $CO_2$ would be the reason for the warm up. I get that! But the multiple shifts of the climate to warm/cold/warm/ cold for the

2000 + years *before* the industrial revolution with no increase in $CO_2$ concentration -- were ignored. Why did science/government officials ignore this fact?

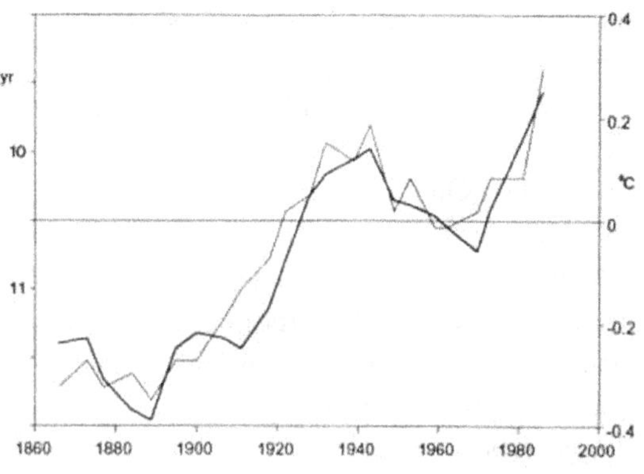


Figure 7.3  Gleissberg record of LSC (see text)

In reality, it became a *perfect political scenario* to ignore the facts; and welcome: increased tax revenue, easy access to science grants, and gains in media advertising – all due to crises claims of the impending natural disasters due to the *postulated* continued warming. The idea that the Sun was involved was quickly put down by the scientists aspiring to gain from the $CO_2$ warming theory – in unison they claimed that the *insolation of the Sun* over the 11-12 year cycle was insignificant to produce the warming.

Later, the published result in *Nature* in 1999, by Lockwood, et al.[8], that the *total magnetic field* leaving the Sun (dragged out by the solar wind) had risen by a factor of 2.3 since 1901 – was an *exceeding strong indicator* that the Sun *was involved*. This fact was also ignore by many – but not by everyone! Many scientists with exceeding strong credentials rose to the occasion and the debate began. You will meet several of these brave souls in the history that follows.

## References: Chapter 7


1.  Neftel, A.H. et al, 1983: Ice core sample measurements give atmospheric $CO_2$ content during the last 40,000 years. *Nature*, **295**, 220-223.
2.  Plimer, I., 2009: *Heaven and Earth -- global warming: the missing science*. Quartet Books, Limited, London, 503 pp.
3.  The Keeling $CO_2$ record from Mauna Loa provided by NOAA/$CO_2$ data/ full record.
4.  Callendar, G. S., 1958: On the amount of carbon dioxide in the atmosphere. *Tellus*, **10**, 243-248.
5.  Akasofu, S., 2009: Two natural components of the recent climate change. Dir. of Intl. Arctic Research Center. http//people.iarc.uaf.edu/~akasofu/pdf/two_natural_components_recent_climate_change.pdf.
6.  Gleissberg, W., 1965: The eighty-year solar cycle in aurorae frequency numbers. *J. Br. Astron. Assoc.* **75**, 227-231.
7.  Landscheidt, T., 2003: New little ice age instead of global warming. *Energy and Environment*, **14**, 327-350. Pointing out the disagreement of the $CO_2$ record with data from Friis-Christensen, E. and K. Lassen, 1991: Length of the solar cycle: an indicator of solar activity closely associated with climate. *Science*, **254**, 698-700.
8.  Lockwood, R., R. Stamper and W. M. Wild, 1999: A doubling of the sun's coronal magnetic field during the past 100 years. *Nature*, **399**, 437-439.

# The Debate Builds Despite the Evidence

The above information had been conveniently ignored and the concept of "Greenhouse gases" causing warming" has been pushed hard by governments for the extra taxes that are applied to the entire fossil fuel industry. The term "greenhouse" originated in the time of Arrhenius. The United Nations created the Intergovernmental Panel on Climate Change (IPCC) in **1988** and that organization kept the term. The term was not acceptable to all and for several reasons!

The term does not apply to the atmosphere as stated by Fleagle and Businger[1] who first objected to Arrhenius' use of the term in their textbook in 1963 – "the greenhouse achieves much higher temperatures than the surrounding air because the glass cover of the greenhouse prevents the warm air from rising and removing heat from the greenhouse."

The atmosphere has no roof, convection is not restricted, in fact convection is *enhanced* is by the lowest thermal layer – a "protective thermal blanket" implied by the above authors, though this is not a perfect term either as a "blanket" would also suppress convection beneath its surface. The role of convection and radiation are described and quantified in Chapter 10.

The atmospheric temperature appears oblivious to $CO_2$ values. Could there be something unique about the role of $CO_2$ in the radiation process? This will be examined in Chapter 11 to see if there is something special about the role of $CO_2$ -- to see if the theory is wrong – and if so – why.

The government/media push for the $CO_2$ as the cause of climate change was not met with universal acceptance – in fact there was and is still vehement opposition to this theory – even before the second climate change theory of Svensmark appeared on the scene. As part of our time travel, we pick up the status of the $CO_2$ climate warming theory in **1989** as expressed in Forbes magazine[2].

On November 7, 1989 the USA and Japan shocked the world's environmentalists by not signing a draft resolution at a world climate conference that would restrict all countries to achieve a particular target of $CO_2$ by 2000. The conference was in the Netherlands and the USA decision was based on the President's scientific advisor, D. Allan Bromley's testimony to Senator Al Gore's Subcommittee on Science, Technology & Space which was: "My belief is that we should not move forward on major programs until we have a reasonable understanding of the scientific and economic consequences of those programs". Japan and 30 other developing nations agreed with the USA position and did not sign the resolution.

This was the beginning of the opposition concerning the reality of the $CO_2$ theory and its stated climatic impact -- what was then known about these two features *(the intensity* of *climate-change* and the *impact* of that change) *versus* what was being projected by the environmental community, the media, and self-serving politicians. The latter groups certainly won the day from this point forward for the next 20 years – until a rival organization was formed and went public in 2009. However, several of the tactics used to win the day were thinly veiled – nature and common sense eventually shifted the tide – but let's keep on the time journey and let the results unfold.

The same article in Forbes magazine used the headline "Apocalypse sells well in the media and on capitol hill." Just a few of the main points expressed in the article are presented here. It was revealed that eleven months earlier *Time* magazine had a cover story on environmental catastrophes, declaring that greenhouse gases would create a *climate calamity.*

One month previous the *New York Times* had a story about how melting polar ice would flood the nations that can least afford to defend themselves – third world countries like Bangladesh and India; and that ads had appeared for Stephen Schneider's book *Global Warming*, with an comment from Senator Tim Worth (D-Colo.), that "painted the picture of seas surging across the land, famine on an epidemic scale and eco-system collapse".

The article also cautioned on not rushing into destroying the world's economies to contain $CO_2$ build-up in view of previous scares: nuclear winter, cancer-causing cranberries and $ 100 oil. Also added to list of "clamitarians" was Stephen Schneider with his then 1976 book, *The Genesis Strategy*, where he supported the then popular view that we were in for a new ice age, "perhaps akin to the Little Ice Age of 1500-1850. Climate variability which is the bane of reliable food production, can be expected to increase along with the cooling."

The same article had a counter example of caution provided by Patrick Michaels of the U. of Virginia. He used a formula provided by one of the then leading climate models predicting that a 1% annual rise in $CO_2$ would create a 0.7 degree centigrade warming in 30 years. Prof. Michaels applied the formula to the period 1950 to 1988, where greenhouse gases rose 1.2% per year. Over this period he found an actual 0.2 degree warming in land temperatures, where the model would have predicted 1.3 degrees. His conclusion was that if a model *cannot predict the past,* being off by 500%, it cannot predict the *future.*

Another magazine article in April of **1990** discussed the "Global Warming's Heated Debate" and raised the intensity question, but primarily focused on the cost of trying to *solve* the problem of $CO_2$ reduction and the impact of that cost on all the other things the country needed to do for society[3].

That same article mentioned a controversial quote from Stephen Schneider that appeared in *Discover* magazine the previous October.

He stated that scientists should consider stretching the truth – his quote "to get some broad-based support, to capture the public's imagination. That, of course, entails getting loads of media coverage. So we have to offer up scary scenarios, make simplified, dramatic statements and make little mention of any doubts we might have… Each of us has to decide what the right balance is between being effective and being honest." This quote did not set well with many scientists, but ample evidence exists that these views were carried out.

A final statement in this article was a summary of the views of a writer and philosopher of science, Alston Chase. Quoting the author who conducted the interview with Mr. Chase, "he saw a disturbing pattern in the way public policy is formed on environmental issues. It begins with environmental groups making claims – often premature and sensational, that represent *one part* of scientific opinion. The media pick them up and a bandwagon develops, which sweeps many grant-seeking scientists along with it. That same bandwagon can also sweep the nation into costly commitments made not in the light of reason but in the heat of politics".

One of first prominent scientists in the USA to become opposed to the $CO_2$ warming theory, as it was presented in **1990**, was the leading atmospheric physicist Prof. Richard S. Lindzen. His American Meteorological Society (AMS) paper[4] of that year showed concern on several points (I will mention just three here): the unwarranted enthusiasm for the $CO_2$ warming theory, the fact that several sources had advocated that *skepticism be stifled,* and that while all the climate models at that time were calling for the greatest warming to be in the polar regions, he included a figure from Rogers[5] that indicated the Atlantic Polar region had been cooling from 1940 through to 1980.

[This matched the two different records of cooling over the 1940 to 1975 period discussed earlier in Chapter 7 on the 20th century cooling.]

This was the beginning of the opposition tagging Lindzen and others with the term "denier" (a name attached to all those who would dispute *any* of the $CO_2$ *warming-science or calamity claims*).

This was a most disgusting term and will not be used further in this book – Lindzen, was and is, one of the most intellectual people I have met and was logically seeking the truth on this climate issue – he knew the dynamics of the atmosphere extremely well and saw through the faulty logic.

Most critics of either point, the *seriousness of the climate issue* or the details of the *various calamites proposed* as a result of the climate change, were experts in their respective fields and knew what they were talking about. We will meet Lindzen again, but he has never left his intellectual position on this matter – in fact he has strengthen his resistance to the 'consensus' view that dangerous human-caused global warming is upon us.

The bandwagon rolled on and the litany of calamities that would befall the world due to the warming climate continued to grow into the early 90's. However, during that period a positive event cheered this author due to a meeting with a very respectable international scientist, Prof. Aksel C. Wiin-Nielsen (my main professor when I received my Ph.D. from the U. of Michigan in 1970).

Prof. Wiin-Nielsen was recruited from being the head of the Department at Michigan to be the first Director of the European Centre for Medium range Weather Forecasts (ECMWF). This organization was the first of its kind and was created in 1973. He started as Director in January 1974, and I visited him during his first week in the office then when his staff was a single secretary. He was an outstanding well-liked scientist, a gentleman, and a wonderful recruiter -- he assembled a great staff and ECMWF has a wonderful record of accomplishment.

In 1979 the 8th World Meteorological Congress appointed him to be the United Nations' World Meteorological Organization's third Secretary-General; he left ECMWF at the end of that year and served as the SG from 1 January 1980 to 31 December 1983. His 4-year term was short by previous standards and he confided to this author that part of his being out-voted by another worthy opponent was his stand against the cause of climate warming not being $CO_2$ which was being pushed hard by the UN.

Aksel was one of the world's experts on the energetics of the atmosphere (published more than 100 peer-reviewed articles and a text book on this and related subjects). He knew the physics involved in the atmosphere as well as anyone. I had the occasion to visit him in the early 90's in Denmark, the country of his birth, where he was by then semi-retired.

We had communicated through the years and he had access to a climate model result that he wanted to share with me. The *best available climate model was used to* perform a 20-year forecast over Western Europe. After considerable checking of long term records, it was decided to select a 20-year period where the temperature actually *did increase* in that part of the world.

The *actual* temperature increases at 8 stations from England, France, Germany, and others, were compared with the climate model predictions for the same stations. Every station temperature increase (all in degrees centigrade) were *over predicted* by the climate model: the ratio of over prediction varied from 2.80 for the best forecast to 10.71 for the worst – the average for the 8 stations was 4.88. This was very poor, but in line with what Pat Michaels had estimated and what Aksel and I had expected from what we knew about the climate models. This revelation was a confirming sign for this author to seek the fundamental proof about the $CO_2$ climate theory.


## References: Chapter 8.1


1.  Fleagle, R. G. and J. A. Businger, 1963: *An introduction to atmospheric physics.* Academic Press, New York, 346 pp.
2.  Brookes, W. T., 1989: The global warming panic. *Forbes*, Dec 25, 1989, 97-102.
3.  Lochhead, C., 1900: Global warming's heated debate. *Insight* magazine, April 10, 1990, 8-18.

4. Lindzen, R. S., 1990: Some coolness concerning global warming. *Bull. Am. Meteorol. Soc.*, **71**, 288-299.

5. Rogers, J. C., 1989: I Proceedings of the thirteenth annual climate diagnostics workshop. P170 (prepared by the national Oceanic and Atmospheric Administration, and available from National Technical Information Service, Department of Commerce, Sills Building,5285 Port Royal road, Springfield, VA 22161).

## CHAPTER 8.2

# CO$_2$ Climate Theory Becomes a Political Tool

The year 1992 was a very bad year! The decade of 1990 to 2000 was very harmful for the science of *climate change* -- it went from a legitimate scientific debate to a political farce. It wasn't until 2009 that things got back under control. The year of 1992 was symbolic of what was to come! The Framework Convention on Climate Change (FCCC) was a treaty signed at the 1992 Earth Summit in Rio De Janeiro, Brazil. The official delegates were 2400 in number from 170 governments.

There was another simultaneous meeting in Rio organized by the Nongovernmental Organizational Forum – the total number of environmental activists in Rio then approached 20,000[1]. The Nongovernmental Organizations (NGOs) are usually non-profit organizations, independent of governments, which are active in social, environmental and other areas to affect changes. It would be 2009 before a counter group to the IPCC would publish a document that would help set the record straight.

The list of attendees in Rio included many individuals and company representatives with an eye towards *getting rich* by "cap and trade" – a system rumored to be eventually used for reducing carbon emissions that was proven partially responsible for reducing the acid rain problem of the 1980s. The problem in the 1980s was that American plants were emitting clouds of sulfur dioxide, which were returned to Earth as acid rain, damaging lakes, forests and buildings across eastern Canada and the United States. Environmentalists were advocating a command-and-control approach where the government would dictate that scrubbers be installed capable of removing sulfur dioxide from power plant exhausts. The acid rain issue was fairly easily solved.

The premise of cap-and trade is that rather than dictate to industry how to pollute less, come up with a more powerful scheme -- simply impose a cap on emissions – each company starts with a reduced allowance of tons (of pollutants) it can produce – and then decides how to use its allowance. The company then had a number of options: restrict output, buy a scrubber, switch to a cheaper fuel, or if it did not use up its allowance, it might sell its unused portion. On the other hand, the company might have to buy extra allowances on the open market. Each year the caps go down for *all companies* and the *shrinking pool of allowances become much more expensive*[2].

This was the background when the IPCC's *Climate Change 1995* was reviewed by its consulting scientists in late **1995**. However, important changes were made to what the scientists had previously concluded. Sir John Houghton, Chairman of the IPCC working group had received a letter from the U.S. State Department dated November 15, 1995. It said:

"It is essential that the chapters not be finalized prior to the completion of the discussion at the IPCC Working Group I plenary in Madrid, and that chapter authors be prevailed upon to modify their text in an appropriate manner following the discussion in Madrid."

This letter was sent by a senior career Foreign Service officer, Mr. Mount, then acting as Deputy Assistant Secretary of State[1]. The Under Secretary of State for Global affairs at that time was former Senator Tim Wirth (D-CO) – who was an ardent supporter of man-made warming, and a close political ally of then-President Bill Clinton and then-Vice President Al Gore. There seems little doubt that the letter that was sent by the Foreign Service officer was directed to be sent by Under Secretary Wirth.[Mount was later named Ambassador to

Iceland – a plum position in a peaceful country – such an ambassadorship was often given to a political ally of the White House[1].]

The Madrid Plenary held in November 1995 was a political meeting with representatives of ninety –six countries and fourteen NGOs. They went over the text of the "accepted" report line by line. The final report was rewritten to reflect the global warming campaign of the UN, the NGOs and the Clinton administration.

The Summary for Policymakers was approved in December and the full report including the important Chapter 8 was accepted. However, the printed report was not released until May of 1996, it was then that the scientific reviewers found that major changes had been made "in the back room" after they had signed off on the science in Chapter 8's contents[1].

Only one statement from the *final printed report* needs to be provided here: "The body of statistical evidence in Chapter 8, when examined in the context of our physical understanding of the climate system, now points to a discernable human influence on the global climate[3]".

The book of Singer and Avery[1] provides several of the key statements that were deleted from the expert-approved Chapter 8 draft. Only one of these is sufficient here:

"Any claims of positive detection and attribution of significant climate change are likely to remain controversial until uncertainties in the total natural variability of the climate system are reduced."

The scientific journal *Nature* mildly rebuked the IPCC for redoing Chapter 8 to "ensure that it conformed" with the report's politically correct Summary for Policymakers. However, the Wall Street Journal was outraged. Its condemning editorial, "Cover-up in the Greenhouse", appeared 11 June, 1996[1]. The next day, Frederick Seitz, former president of the National Academy of Sciences, detailed the illegitimate rewrite with a commentary titled, "Major Deception on Global Warming[4].

The Petition Project[5] was a campaign to counter the political spin of the climate warming issue that took place in the mid-90s; and to state unequivocally that there was no scientific "consensus" about the cause of the climate warming. It was initiated in **1998** by a group of scientists with a campaign that gathered thousands of signatures during 1998 – 1999. Between 1998 and 2007 the list of petition signatories grew gradually, without a special campaign. Between October 2007 and March 2008 a new campaign for signatures was initiated.

Opponents of the Petition Project would sometimes submit forged signatures to discredit the project. These were found and removed. The Petition was signed by 31,478 Americans with university degrees in science, including 9,029 with Ph.Ds[5]. The Petition states that "there is no convincing scientific evidence that human release of carbon dioxide, methane, or other greenhouse gases is causing, or will in the foreseeable future, cause catastrophic heating of the Earth's atmosphere and disruption of the Earth's climate. Moreover, there is substantial scientific evidence that increases in atmospheric carbon dioxide produce many beneficial effects upon the natural plant and animal environments of the Earth".

Michael Crichton has since passed away (1942 – 2008), but he was a very productive author (his books selling over 200 million worldwide), film director and producer (you may have seen *The Andromeda Strain* and *Jurassic Park* – two of 12 books which were made into films), and he was the creator/ writer/ executive producer of the television series *ER*. His basic genre of literature was the techno-thriller and in **2004** he published *State of Fear* – where his "bad guys" were radical environmentalists trying to scare the world about global warming in order to make big money.

That brief climate background got Crichton (age 64) and Prof. Richard Lindzen Age 67) as a duo on a debate team against three scientists pushing the dangerous human-causing global warming. The debate was held in **2007** in New York with the title "Global Warming Is Not a Crisis" staged by a group called Intelligence Squared U.S.[6] The scientific opposition team was headed by Richard Somerville (age 64), U. of California and San Diego climatologist.

Somerville attacked the "not a crisis" position."[A crisis] does not mean catastrophe or alarmism", he stated "it means a crucial or decisive moment, a turning point, a state of affairs in which a deceive change for better or worse is eminent. Our task tonight is to persuade you that global warming is a crisis in exactly that sense. The science warns us that continuing to fuel the world using present technology will bring dangerous and possibly surprising climate changes by the end of this century, if not sooner."

Crichton insisted that pressing real-time problems are far more important than an iffy, long term one. "Every day 30,000 people on this planet die of the diseases of poverty," he tells the crowd. "A third of the planet doesn't have electricity. We have a billion people with no clean water. We have half a billion people going to bed hungry every night. Do we care about this? It seems that we don't. It seems we would rather look a 100 years into the future than pay attention to what is going on now." The debate was won by the Crichton and Lindzen team by audience vote.

The writer describing the debate scenario was on the side of the "human caused global warming" and added further comments along those lines, complaining that the debate did not cover the degree of $CO_2$ warming. However, the winning team would have easily won that side also as, at that point in time, both realized that any further doubling of $CO_2$ would lead to a trivial rise in temperature.

Al Gore's movie *An Inconvenient Truth* was a Hollywood produced documentary movie made in **2006** that was well received by the public and well-attended. It received an Academy Award and Mr. Gore made lots of money. However it was full of alarmism and fundamental errors. There were 35 errors in the movie as found by Monckton[7], but the judge in a High Court in London in October of **2007** had not time to consider more than 9 of those errors. The judge found those *9 errors serious enough t*o require the UK government to pay substantial costs to the plaintiff in a law suit.

The judge had stated that, if the UK Government had not agreed to send to every secondary school in England *a corrected guidance note* making clear the mainstream scientific position on these nine "errors", he would have made a finding that the Government's distribution of the film and the first draft guidance note earlier in 2007 to all English secondary schools had been an unlawful contravention of an Act of Parliament *prohibiting the political indoctrination of children.*

Only 3 of those 35 errors are mentioned here, but the reader can read about all of them in the Monckton reference above. Sea level rise was to be as high as 6 meters (~20 feet) as stated by Gore, but the IPCC itself estimate is 6 centimeters over the next century – thus Gore exaggerated by a factor of 100. Gore clearly stated that $CO_2$ was driving temperature, but the studies of Ice Core data in Chapter 3 indicate the exact opposite – three different scientific missions by three different countries found that temperature leads the $CO_2$ by several 100 years over the past 420,000 years.

A final error discussed here of those in Gore's movie is the statement that hurricane Katrina, which devastated New Orleans in 2005, was caused by global warming. It was not, it was caused by the failure of the administration of New Orleans that after 30 years of warning by the Corps of Engineers that the levees that kept New Orleans dry could not stand a direct hit by a hurricane.

Katrina was only a category 3 storm when it struck the levees – they failed as the Corps said they would – the Democratic Party running the city with no action was to blame, not global warming.

The number of Atlantic hurricanes shows no trend over the past 50 years; the number of typhoons has fallen over the past 30 years; the number of tornadoes has risen because of better detection systems for smaller tornadoes; but the number of larger tornadoes in the U.S. has fallen[7]. For the first year since records were kept in 1950, the have been no violent tornadoes[8] in the U.S. in 2018 – this means no EF4 or EF5 ratings on a scale of 0 to 5. [Weather is chaotic, this probably will not happen two years in a row.]

The evils of and miss-uses of 'cap and trade' became evident after several years of implementation and with no major changes in the environment having occurred as had been predicted. The *Financial*

*Times* out of London had the following comments in an article dated April 25, 2007 concerning the carbon offsetting industry.

"Companies and individuals rushing to green have been spending millions on "carbon credit" projects that yield few if any environmental benefits. A *Financial Times* investigation has uncovered widespread failings in the new market for greenhouse gases, suggesting that some organizations are paying for emission reductions that do not take place. Others are meanwhile making big profits from carbon trading for very small expenditure and in some cases for clean-ups that would have been made anyway."

The same article goes on to say: "The growing political salience of environmental politics has sparked a green gold rush, which has a dramatic expansion in the number of businesses offering both companies and individuals the chance to go "carbon neutral". Offsetting their own energy use by buying carbon credits that cancel out their contribution to global warming."

The *Financial Times* investigation found: (1) widespread instances of people and organizations buying worthless credits that do not yield any reductions in carbon emissions, (2) industrial companies profiting from doing very little -- or from gaining carbon credits on the basis of efficiency gains from which they have already benefitted substantially, (3) brokers were providing services of questionable or no value, (4) a shortage of verification, making it difficult for buyers to assess the true value of carbon credits, (5) companies and individuals being charged over the odds for the private purchase of European Union carbon permits that have plummeted in value because they do not result in emission cuts.

The second edition of the book by Singer and Avery[1] made a big impact in **2008**. It became a *New York Times* best-seller and stayed on that distinguished list for several months. It offered an opposite view and counter scientific data that competed with Al Gore's false claim of the $CO_2$ role in *climate-change* – introduced without any proof; and the even worse shameless exaggeration of what additional global warming would produce in terms of violent environmental results.

Gore's campaign was indeed aided by his infamous movie, and by environmental advocacy groups, government agencies, and the media having spared no expense in promoting his message.

The book of the Australian geologist Ian Plimer[9] in **2009** was also a tremendous counter to the alarmist crowd of the $CO_2$ climate theory. He has received many awards and his previous book, *A Short History of Planet Earth,* won the Eureka Prize. His detail about past *climate-change* -- with both warm and cold regimes is outstanding.

Finally, the best answer to the politicization of the climate issue came with the creation of the Nongovernmental International Panel on Climate Change (NIPPC) which was set up to examine the same data used by the United Nation's IPCC. This provided the world with a second opinion on the important issues. Their report in **2009** disagreed with the IPCC result that the climate was *very likely* caused by anthropogenic greenhouse gas emissions – it was stated that the man-made emissions of greenhouse gases were *not playing a substantial role,* rather it was natural causes that were the dominant cause of *climate-change*[10].

A second important difference between the two organizational results was that the IPCC stated that global warming will "increase the number of people suffering from death, disease and injury from heatwaves, floods, storms, fires and droughts." The NIPCC conclusion was the opposite "a warmer world will be a safer and healthier world for humans and wildlife alike."

A very convincing statement within the NIPCC report of why the IPCC reports are marred with controversy and frequently contradicted by subsequent research is because "its agenda to find evidence of a human role in climate change is a major reason; its organization as a government entity beholden to political agenda is another reason; and the large professional and financial rewards that go to scientists and bureaucrats who are willing to bend the scientific facts to match those agendas is yet a third major reason."

The UK has apparently made a huge error in **2008** -- not following the warning of Alston Chase – "swept up by the bandwagon into costly commitments made not in the light of reason but in the heat of politics". The UK's Climate Change Act of 2008 required reduced carbon emissions by 80% by 2050. The government was to set up legally-binding 'carbon budgets" – every 5-years to act as stepping stones toward the 2050 target. The carbon budget is a cap on the amount of greenhouse gases emitted in the UK over a five-year period.

Matt Ridley, a British journalist and businessman, created quite a stir with his presentation of the Angus Millar Lecture of the Royal Society of the Arts, in Edinburgh, 31 October **2011**. He pointed out that the climate did change naturally in the past (without mankind's $CO_2$ influence) – stalagmites, tree rings and ice cores all confirm that it was significantly warmer 7000 years ago. He pointed out that sea level is rising at the *'unthreatening rate of a foot per century and is decelerating'* Greenland is losing ice at the rate of 150 gigatonnes per year, which is 0.6% *per century*. Tropical storm intensity and frequency have *'gone down, not up, in the last 20 years*'[11].

He added "remember Jim Hansen of NASA told us in 1998 to expect 2-4 degrees of warming in 25 years. We are experiencing one-tenth of that." He defined himself a 'heretic' for no longer accepting what had been preached on climate change, and called the $CO_2$ warming enthusiasts 'alarmists'. Probably his most impactful statement was that "the alarmists have been handed power over our lives; the heretics have not. Remember Britain's unilateral Climate Change Act is officially expected to cost the hard-pressed UK economy Ł18.3 billion a year for the next 39 years and achieve an unmeasurable small change in carbon dioxide levels".

The rapid dissemination of this speech probably helped the next action to occur – four months later in February of **2012**, Prof Lindzen was invited to the British House of Commons to lecture on the subject of *climate-change*, as the politicians were reconsidering the Climate Change Act. The main points of Lindzen were: the evidence is that the increase in $CO_2$ will lead to very little warming, and the connection of this minimal warming (or even significant warming) to the purported catastrophes is also minimal; "the arguments on which the catastrophic claims are made are extremely weak – and commonly acknowledged as such; they are sometimes overtly dishonest".

An important article appeared in the Daily Telegraph in the UK in December **2015** by Owen Patterson[12], a former MP. He was quite disturbed with the Climate Change Act. Just three of his points are reproduced here which match the warnings of Alston Chase. These are all direct quotes from his article. "Apart from Britain, we are left uniquely isolated and vulnerable as the only country in the world with a *legal* target for reducing emissions, thanks to our Climate Change Act of 2008. No other country will be breaking its own law if it misses its target. But we have a binding target to reduce emissions by 80 per cent by 2050. We have repeatedly boasted that we are setting the world an example – but the world seems disinclined to take notice. Our dash for wind power so distorted the electricity market that it has actually prevented the construction of efficient and cheap combined-cycle gas turbines."

"The 2050 target commits us to decarbonizing our electricity, abolishing gas as a fuel for cooking and heating our homes, and converting two thirds of our cars to electric. These aims come at an astronomical cost. Since wind does not significantly reduce emissions (because of the need for backup when it is not blowing) and because solar power is useless at night and in winter, it would mean a vast investment in nuclear power, equivalent to building a new Hinckley Point every three years for 35 years, That's neither feasible nor affordable."

It was clear that the idea of total abandonment of fossil fuel use was never carefully evaluated by the alarmists. It was not physically possible to do this in any realistic time period. [An update on world progress on this fossil fuel point is available in Chapter 14.] The shear amount of land required for solar, wind, and bio-fuels would have devoured a very significant part of land for food production, and removed land from forest growth. No attention was really dedicated to determining the price for this conversion – and the subsequent impact on the economy.

The barrage of warming hype from the media eventually defeats itself; smart people begin to look at the facts and raise questions. After seven years of the Protocol being essentially inactive (from 2005 to 2012), there was very little warming: sea levels were not rising faster, there were no species extinctions, and death rates were not increasing.

Many people began to see through the constant media rhetoric of doom and gloom, and realized that there were benefits of increased $CO_2$ levels in the atmosphere that would increase crop yields and forest growth. People became aware of the Roman warming and the Medieval Warming which were very productive periods for humankind -- that a warmer climate was good for humankind.

Proof of the above transformation in the minds of those who use their intellect to think things through, we take the reader back to the 'Copenhagen Consensus' of 2004. This was described in the book of Singer and Avery[1].

The Copenhagen Consensus is a panel of the world's leading economists formed to propose the most effective way to use $50 billion to benefit mankind. "In 2004 the top four recommendations for the use of the money were: (1) for combating new cases of AIDS, (2) reducing iron deficiency anemia in Third World women and children, (3) controlling malaria that kills 2.7 million annually, and (4) on agriculture research to sustainably raise crop yields and ease the competition for land between people and wildlife".

The Kyoto Protocol was a UN effort to extend nations commitments to cut $CO_2$ by the year 2005. The Copenhagen Consensus ranked the Kyoto Protocol sixteenth out of seventeen proposed ways to use the money. The economists said that the Kyoto's costs outweigh the benefits – even though the *$CO_2$ warming was assumed to be real.* Had they known then what we know now, that $CO_2$ has no role in climate change, the Kyoto Protocol would not have been on the agenda at all.

# References: Chapter 8.2


1.  Singer, S. F. and D. T. Avery, 2008: *Unstoppable Global Warming.* Rowman &Littlefield Publishers, Inc., 278 pp.
2.  Conniff, R., 2009: The political history of cap and trade. *Smithsonian Magazine, August, 2009.*
3.  Intergovernmental Panel on Climate Change (IPCC); Climate Change 1995, Chapter 8, p.429.
4.  Seitz, F., 1996: Major deception on global warming. *Wall Street Journal,* 12 June, 1996.
5.  Singer, S. F. and C. D. Idso, 2009: *Climate Change Reconsidered.* The report of the Nongovernmental International panel on Climate Change. The Heartland Institute. 856 pp.
6.  Revkin, A., 2007: Global meltdown. (*New York Times* writer), AARP magazine July-August.
7.  Monckton, C. 2997: 35 inconvenient truths. Science and Public Policy Institute. http;//ben-israel.rutgers.edu/711/Monckton-response-to –gore-errors.pdf.
8.  https://www.thegwpf.com/2018-will-be-the-first-year-with-no-violent-tornadoes-in-the-u-s/
9.  Plimer, I., 2009: *Heaven and Earth -- global warming: the missing science.* Quartet Books, Limited, London, 503 pp.
10. Singer, S. F. and C. D. Idso, 2009: *Climate Change Reconsidered.* The report of the Nongovernmental International panel on Climate Change. The Heartland Institute. 856 pp.
11. Ridely, M., 2011: Scientific heresy. http//www.ratonaloptimist.com/blog/scientific-heresy/.
12. Patterson, O., 2015: Why we should scrap the Climate Change Act. https://www.teegraph.co.uk/new/earth/environment/climate change/12046531/

# The Tide Begins to Turn

Our narrative must go back in time a bit to pick up the early at-tribution of *climate-change* due to astronomical conditions and especially the Sun. It was the incredible Sir Isaac Newton[1], recognized as one of the most influential scientists of any age, who first brought to light that the Sun is not the center of mass of the solar system, but in fact orbits around that center of mass -- called the solar system barycenter (SSB). The SSB is primarily determined by the mass of the Sun and the four largest planets within the solar system. Newton came to this conclusion not by observation, but by his use of his own mathematical derivation of the laws of gravitation.

The application of the Sun's motion about the SSB was brought to attention first by the Australian scientist Rhodes Fairbridge. This prolific writer contributed input for many scientific disciplines, and authored or edited more than 100 scientific books. His mention of the SSB and the possible impact on climate was in 1961[2]. Through-out his long career (he died at the age of 92) be showed evidence of climate change in glaciations, sediments, and the records of the iso-topes of carbon, beryllium, oxygen, chlorine and in tree rings and ice cores.

Fairbridge insisted that the solar variability be examined entirely, and that once this was accomplished it would reveal that solar varia-bility on the Earth's climate would be strongly *nonlinear, stochastic and significant*[3] (all three of these attributes have been found to be absolutely true). The details of the Sun's motion about the SSB is far more complete today than in his time. Extremely accurate data is presented in Chapter 12. Nevertheless, the pioneering solar work of Fairbridge prompted others to explore the solar influence on climate change.

Various scientists have challenged the IPCC's assessment of the impact on surface temperature increase with a doubling of $CO_2$ with various claims of 2.5 to 5°C; Manabe and Weatherald[4] said 2.9°C; Sorokhtin, et al[5] said less than 0.02 to 0.03°C in 2007.

Having been impressed with the Russian's book (Sorokhtin, et al.) this author wrote a short paper pointing out their new approach (see Appendix C). The basis of the author's paper were the many results of the three Russian authors. That paper was submitted in 2008 and it was immediately rejected by the first reviewer (who sided with the IPCC) and *admitted in his review* that he did not even bother to read the Russian's book! Such was the plight of many scientists in those years who did not follow the claims of the IPCC.

Plimer[6] in 2009 recorded his views that $CO_2$ in the atmosphere acts much like a curtain on a window – add one to keep the light out and adding more had little effect. This is why he claimed that even in periods of climate change with as much as 25 times the current value of $CO_2$ concentration in the past, there was no correlation of $CO_2$ within the Ice Ages.

The solution shown in Chapter 11 of this book *indicates no impact of $CO_2$ in changing the climate* – as water vapor and $CO_2$, while both contributing to the low level thermal blanket, provide no net heating to the atmospheric column. Doubling $CO_2$ has no impact – Chapter 11 shows the $CO_2$ *density change with height matters* -- this is very small because both pressure and temperature are involved; the *Planck effect with temperature decreasing dominates.* Further, whatever heat is available at the surface is absorbed and then re-emitted upward, combining with the other two physical forces of convection and latent heat release, to balance the net radiation for the planet as a whole. This subject of the three forces is discussed in detail in Chapter 10.

Dr. Habibullo Abdussamatov received his doctorate at the University of Leningrad. His main interest was the total solar irradiance (TSI). While there is an 11-year cycle of sunspots and TSI, the variance of TSI is too small to correlate with climate change. Complicating that correlation, was his view that ocean heat is released some 15-20 years later after the irradiance has subsided. The length of this delay is considerable longer than estimates from other scientists. Abdussamatov speaks of another 200 – year cycle of TSI and believed in 2007 that the total irradiance was about to fall reaching a minimum around 2041 +/- 11 years.

He is a non-believer in the atmosphere's greenhouse effect and stated (perhaps "tongue-in cheek") that "Heated greenhouse gases, which become lighter as a result of expansion, ascend to the atmosphere only to give the absorbed heat away"[7]. This is not the proper scientific explanation which will be given in Chapters 10 and 11, but is a less complicated way of looking at the issue.

Abdussamatov made the "The Deniers" list of Solomon[8] which included "world - renowned scientists (Prof. Lindzen was another) who stood up against global warming hysteria, political persecution, and fraud". Famous people made this list, and they were "applauded for their credentials that were often far more impressive than those of some of the gurus propounding climate-change catastrophes". While the author of this book made that statement, he concluded at the end of his book that the group failed to convince him that "global warming was all a hoax". The author was wrong on three counts: they agreed global warming was real, they believed that the $CO_2$ cause was a hoax, and he was wrong in setting himself up as smarter than his interviewed subjects.

Nicola Scafetta received his Ph. D. in physics in 2001. He was skeptical of man-made climate forcing and stated his work on solar forcing with TSI. In his paper of 2007 with West[9] they did not use TSI reconstruction as radiative forcing, but as a proxy for the *entire solar dynamics* (UV, cosmic rays, magnetic fields, etc.) They concluded that solar forcing significantly altered the climate system with a slow response time of 6 to 12 years due to the thermal inertia of the oceans.

In his 2010 paper Scafetta[10] found climate change oscillations with peak-to trough amplitude of about 0.1 °C and 0.5 °C, and periods of about 20 and 60 years respectively, that were synchronized to the orbital periods of Jupiter and Saturn. More specifically, the orbit of the Sun around the SSB can be easily evaluated using the NASA Jet propulsion Laboratory Developmental Ephemeris. Both the distance of the Sun relative to the SSB, and the speed of the Sun relative to the SSB can be accurately determined. This will be addressed more thoroughly in Chapter 12.

Francois Gervais is a retired and very well respected physicist from France with a long list of publications and a reputation of concise statements that get to the point in question. His 2014 paper[11] concluded that the evidence for the impact of anthropogenic $CO_2$ was so "trivial and tiny that it was not inconsistent with a null hypotheses of its cause of climate change".

Two years later, after further research, his 2016 paper[12] considered several natural cycles and the natural 60-year cycle due to the correlation of the velocity of the motion from the Sun with respect to the SSB – to be sufficient to question the "dangerous anthropogenic climate warming". He went further: "On inspection of a risk of anthropogenic warming thus toned down, a change of paradigm which highlights a benefit for mankind related to the increase in plant feeding and crops yields by enhanced $CO_2$ photosynthesis is suggested".

This author, Fleming[13] published a peer-reviewed paper in March of **2018**. The results of this paper were first presented to the public in a Keynote address for the 5th World Conference on Climate Change and Global Warming (May 23-24, 2018) in New York City USA. This paper pointed out the proof of $CO_2$ not causing climate change – this from the observational data from history, past and present, which showed zero correlation in climate regimes be they warm or cold. Further mathematical proof that $CO_2$ does not cause *climate-change* proof was presented in that paper, and this proof has been enhanced in the text of Chapter 11 and in Appendix D.

## References: Chapter 8.3


1. Newton, I., 1687: Mathematical Principles of Natural Philosophy. A New Translation by I. Bernard Cohen and Anne Whitman assisted by Julia Budenz. Preceded by A Guide to Newton's Principia by I. Bernard Cohan. University of California Press USA 1999.
2. Fairbridge, R. W., 1961: Convergence of evidence on climate change and ice ages. In *Physics and Chemistry of the Earth* **4**, Pergamon Press, London, 542- 579
3. Mackey, R., 2007: Rhodes Fairbridge and the idea that the solar system regulates the Earth's climate. Journal of Coastal Research, Special Issue 50, 955 – 968.
4. Manabe, S. and R. T. Weatherald, 1967: Thermal equilibrium of the atmosphere with a given distribution of relative humidity. *J Atmos. Sci.*, **24**, 241-259.
5. Sorokhtin, O. G., G. V. Chilingar and L. F. Khilyuk, 2007: *Global warming and global cooling: evolution of climate on Earth*. Elsevier, Amsterdam, 313 pp.
6. Plimer, I., 2009: *Heaven and Earth -- global warming: the missing science*. Quartet Books, Limited, London, 503 pp.
7. Abdussamatov, H. I., 2007: Look to Mars for truth about global warming. https//web.archive.org/web/20070306233723/http://www.canada.com/nationalpost/story.html.
8. Solomon, L., 2008: The Deniers. Richard Vigilante Books, 239 pp.
9. Scafetta, N. and B. J. West, 2007: Phenomenological reconstruction of the solar signature in the Northern Hemisphere surface temperature records since 1600. *JGR*, **112**, 1-10.

10. Scafetta, N., 2010: Empirical evidence for a celestial origin of the climate oscillations and its implications. *J. Atm. And Solar-Terr. Phys.*, **72,** 1-14.

11. Gervais, F., 2014: Tiny warming of residual anthropogenic $CO_2$. *International Journal of Modern Physics B,* **28**, *1-20.*

12. Gervais, F., 2016: Anthropogenic $CO_2$ warming challenged by 60-year cycle. *Earth Science Reviews,* **155**, 129-135.

13. Fleming, R. J., 2018: An updated review about carbon dioxide and climate change. *Environmental Earth Sciences,* **77,** 1-13.

# CHAPTER 9

# The Irrigation System of Planet Earth

Nearly every evening (when no clouds are present) one can marvel at our breathtaking universe. However, *every morning when I awake,* I am gratified and exhilarated by the most important possession of our planet – the *Earth's life giving irrigation system* composed of the atmosphere and oceans circulation systems, the rivers, lakes, the lunar tides which daily clean our shores and, yes, even the clouds that provide a spectacular array of color and form.

The large scale instability of the atmosphere, which was discussed in Chapter 4, and the world's oceans which cover 71% of the Earth's surface, play a major role in *the irrigation system of planet Earth.* But there are other important components! How important is this system?

Imagine a powerful and loving God who created humankind with a reproduction system that combines egg and sperm to evolve into an *adult thinking intelligent being* -- along with a complex system of deoxyribonucleic acid (DNA) – a molecule that contains the instructions an organism needs to develop, live, and reproduce. Such a caring Creator would also provide life-giving water to our Earth with an irrigation system designed to support life on the planet *for a very long time.*

This irrigation system has obviously been important for preserving life on Earth. It is also contributes to the action of the three processes (discussed in Chapter 10) that create the Earth's thermal blanket and subsequent energy balance. The irrigation system is a key feature in different *climate-change* regimes. A quick review of our world's irrigation system is in order.

Water is quite unique, it has molecules that exist in all three phases -- solid, liquid and gas forms that can exist together all at the same time – the only substance on Earth with that capability. The $H_2O$ molecules have the hydrogen atoms attached to the oxygen atoms in such a way that one side of the molecule has a *negative* charge and the opposite side of the molecule has a *positive* charge. This arrangement allows molecules of liquid water to be attracted to each other more easily and form bonds and flow. *The molecular structure of ice* is such that the molecules arrange themselves in a geometric pattern such *that the volume is increased, thus the density is reduced, and ice floats on the top of liquid water.* Thus, marine creatures can survive beneath frozen rivers and lakes, and icebergs float at the ocean surface. The density of water molecules is provided in Table 9.1 from Pidwirny[1]. The maximum density of water occurs at a temperature of 4 °C.

Table 9.1  Density of water in various forms as a function of temperature

| Temperature (degrees Celsius) | Density (g/ cm3 ) |
| --- | --- |
| 0 (solid) | 0.9150 |
| 0 (liquid) | 0.9999 |
| 4 | 1.0000 |
| 20 | 0.9982 |
| 100 | 0.0006 |

The molecular structure of *liquid water* has molecules that are semi-ordered, thus small groups of joined molecules allow the liquid water to flow. The gas phase of water (vaporized water or water vapor commonly referred to the public as *relative humidity*) has molecules that have a random structure with high energy – thus the molecules are less likely to bond together.

Water has another important feature of being a universal *solvent* which is able to dissolve large numbers of different chemical compounds – thus, allowing water to carry solvent nutrients in runoff, groundwater flow and in living organisms. The human body is 60% water. Our blood is ~78% water in which are dissolved or suspended the complex substance that carry on the body's life processes. Water conducts heat easier than any liquid except mercury. This conduction causes large bodies of water (lakes & oceans) to have nearly uniform vertical profiles of temperature[1].

It is estimated that there are approximately 1.360 billion cubic kilometers of water on the planet, primary in liquid form, but only 2.8% of that liquid water is fresh[2]. Water circulates through the hydrological cycle between the atmosphere, biosphere, lithosphere and hydrosphere. The different reservoirs include: the atmosphere, glaciers, groundwater flows, lakes, oceans, rivers, and soils.

In every man-made irrigation system (such as those for a farm or for a golf course) there are four components: a *power system, the water source*, the *distribution and drainage system*, and the *management and control system*. The management and control is very important especially for farmers and greens keepers determining which areas get how much water and when. This *management* must also cope with problems like broken pipes, nature's abrupt changes and so on.

God designed the Earth's irrigation system. He chose the Sun as the *power source*, the *oceans as the primary water source*, the *distribution system* was and is an ensemble of phenomena. The atmosphere's chaotic motion of large scale baroclinic waves assured a random, but sufficiently abundant rainfall for major portions of the continents. Other phenomena providing distribution include the intermittent mid-latitude cyclonic activity, randomized small scale convective storms, orographic uplift over mountains and coastlines, monsoons and sporadic hurricanes. The Earth's water is stored in lakes, rivers, underground reservoirs and of course in our vast oceans.

Having *set up the power source*, the *water source*, and the *distribution and drainage system* for all time, the Creator left the *management and control system* completely up to mankind! His system works pretty well, it has fed humanity for many thousands of years, but it is not perfect with droughts and floods occurring as is His will. *God has made humankind responsible for management and control:* water use in general, irrigation, surface and groundwater control, dams, reservoirs, water quality, wastewater treatment, water laws and treaties! Humankind has partially maintained these responsibilities, but we still have a long way to go – we have much to learn about flood control! Managing water will not get easier with the climate-change to come!

The warm versus cold climate regimes are only somewhat predictable in terms of water availability, but are not consistent over a complete *climate-change* regime. The warmer periods allow the atmosphere to carry more water vapor, but there have been examples of both warm and cold regimes that have had both floods and droughts. One must carefully track the flow of water across our planet. The oceans of Earth cover 71% of the Earth's surface. The world has an enormous amount of water. The distribution of water over land, the sea and underground is estimated in Table 9.2 from the Water Encyclopedia[2].

Table 9.2  World Water Supply Volume (All numbers are in *Thousands* of units)

| Water item | Cubic Miles | Cubic Kilometers | Percent of total water |
|---|---|---|---|
| **Water in Land Areas** | | | |
| Fresh water lakes | 30 | 125 | 0.009 |
| | | | |
| Saline lakes and inland seas | 26 | 104 | 0.008 |
| Rivers (average instantaneous volume) | 0.3 | 1.3 | 0.0001 |
| Soil moisture and vadose water | 16 | 67 | 0.005 |
| Ground water to depth of 4000 meters | 2000 | 8,350 | 0.61 |

| Water item | Cubic Miles | Cubic Kilometers | Percent of total water |
|---|---|---|---|
| Icecaps and glaciers | 7000 | 29,200 | 2.14 |
| **Total in land area (rounded)** | **9,100** | **37,800** | **2.8** |
| | | | |
| **Atmosphere** | **3.1** | **13** | **0.001** |
| **World ocean** | **317,000** | **1,320,000** | **97.3** |
| **Total all items (rounded)** | **326,000** | **1,360,000** | **100** |

The amount of water in the world is overwhelming, more than 1.360 billion cubic kilometers. However, the abundance is misleading as much of the world's water is not available for immediate use. Table 9.2 indicates that 97.3% is available in the oceans, 2.14% is locked up in icecaps and glaciers, and at any one time only a fraction of a percent is fresh water in lakes and rivers.

A major concern for humankind is the distribution of water: making it available at the right place, and in time to meet human needs. The average *daily consumption* per person ranges from 30 liters in the developing countries to 6000 liters in the most developed countries.

The World Meteorological Organization (WMO) reports that it requires approximated 2,500 kilograms of water to produce a loaf of bread and the flour it contains, 17,000 kilograms of water to produce half a kilogram of beef and 290,000 kilograms of water to make a ton of steel[3].

It is obvious that no nation can plan the best uses of its water resources unless it has the properly educated staff and the essential facilities to access and utilize those resources for hydrological (water use and control) forecasting. The economic and social importance hydrological forecasting can be exemplified from the data over Asia in the late 1970s – estimated average annual cost of flood damage was US

$3 billion[3]. One should not attempt to put a value on the human lives lost.

The magnitude of some water runoff processes[4] are indicated in Table 9.3

Table 9.3 Precipitation and various water runoff processes

| Water item | Cubic Miles | Cubic kilometers | Percent of total water |
|---|---|---|---|
| **Annual evaporation (420,000 km² water involved)** | | | |
| From world oceans | 85 | 350 | 0.025 |
| From land areas | 17 | 70 | 0.005 |
| **Total** | 102 | 420 | 0.031 |
| **Annual precipitation** | | | |
| On world oceans | 78 | 320 | 0.024 |
| On land areas | 24 | 100 | 0.007 |
| **Total** | 102 | 420 | 0.031 |
| | | | |
| **Annual runoff to oceans from rivers & icecaps** | 9 | 38 | 0.003 |
| **Ground-water runoff to oceans (5% of surface)** | 0.4 | 1.6 | 0.0001 |
| **Total** | 9.4 | 39.6 | 0.0031 |

Flood control and the related control of water for irrigation are both extremely important. Many ancient civilizations practiced irrigation with canals, dams, dikes and water storage facilities. Today's world cropland in estimated to be 18% fed by a variety of irrigation methods. There is a concern that in times of *climate-change* that the challenge of implementing flood control procedures, combined with efforts to maintain water for irrigation for adequate food production, will not impact the further action required to preserve fresh water for drinking and hygiene.

In the last Chapter of this book there is a discussion of a potential cold period that will replace the Modern Warming that has occurred since 1850. Should this develop as some have predicted, humanity will be severely challenged to meet the fresh water demands of the planet – unless sufficient international plans are put in place prior to that increased demand for fresh water.

One needs to say something about the oceans and sea level. There have been enormous sea level rises and falls with the coming and going of the ice ages. However, since the last glaciation ended 14,000 years ago, the sea level rise has not damaged coral reefs. The major rise ended 6000 years ago and there has been no noted acceleration of sea level rise since the industrial revolution[5].

The oceans have always been alkaline (opposite of acidic) even when the atmospheric $CO_2$ content was 25 to100 times today's values. The chemical interaction of water and $CO_2$ to form carbonate rocks was discussed in the beginning of Chapter 3. The proof of the ocean not becoming acidic from $CO_2$ is supplied by several scientists in Chapter 13.

The oceans are driven by the winds within the atmosphere. The winds are created by pressure gradients within the atmosphere and the rotating Earth. The ocean currents driven by the wind transfer a tremendous amount of heat around the world in various systematic currents. There is a need to study the ocean circulation on a global scale -- *not for the IPCC's definition* of *climate-change;* but for *climate-change* on regional spatial scales and on yearly to decadal time scales.

The magnitude of the ocean transport and the mixing of deep ocean water is still poorly known[5]. Further instrumentation for more ocean observation from the surface to the bottom waters would benefit ocean/atmosphere climate research activities and other scientific applications in all the oceans. There is further suggested research on these matters discussed in Chapter 14.

The entire hydrological cycle will change with the next climate change and humankind will be challenged. Several suggested actions are included within Chapter 15.

## References: Chapter 9


1. Pidwirny, M., 2006: *Physical properties of water.* Fundamentals of Physics, 2$^{nd}$ Edition.
2. van der Leeden, F., F. L. Troise and D. K. Todd, 1990: *The water encyclopedia,* 2$^{nd}$ edition. Lewis Publishers Inc. 808 pp.
3. World Meteorological Organization, 1977: Weather and water. ISBN 92–63-10463-8, 24 pp.
4. U.S. Geological Survey, 1967: National Center, Reston Virginia, USA
5. Plimer, I., 2009: *Heaven and Earth -- global warming: the missing science.* Quartet Books, Limited, London, 503 pp.

CHAPTER **10**

# The Source of the Earth's Thermal Blanket and Energy Balance

The Sun's output has been steady for the last one *billion* years. Three processes have maintained the energy balance over that time period -- any systematic deviation in either direction over such a long period of time would have made life on Earth impossible – a planet too hot (a burnt cinder) or too cold (a ball of ice). These three processes are: radiation, latent heat release from condensation, and atmospheric convection. These all work together to heat the lower atmosphere, transfer heat upward, cool the upper atmosphere and achieve radiative balance.

The Earth receives incoming solar radiation and radiates terrestrial longwave radiation. This radiant energy travels in waves at the speed of light ($3 \times 10^8$ m/s in a vacuum). Details of radiation are described in Chapter 11 and in Appendix D.

Water exists on Earth in three phases – solid, liquid and gas forms. The intermolecular forces in water molecules are *decreased* as energy (heat) is applied to the phases of water. The following water phase changes *require inputs of energy* for changes "upward": ice to liquid (melting or fusion, 334 Joules/ gram), liquid to vapor (evaporation, ~2500 J/g), and ice to vapor (sublimation, ~2834 J/g). Phase changes "downward" *liberate or release similar amounts of*

87

*energy*: liquid to ice (freezing), vapor to liquid (condensation), and vapor to ice (deposition).

The most important phase changes of water for the climate system are the Sun's energy (2500 Joules/gram) evaporating water from the oceans (also from lakes and streams), and then that energy *being released later in the atmosphere as latent heat of condensation.*

Convection is clearly the most important mechanism for transferring heat upward (Emanuel[1]). As hot air expands it becomes less dense and rises. Similarly, denser cooler air drops down and replaces the warmer air. This is a *diabatic* process as there is always some entrainment (*conduction* – molecular collisions exchanging heat – molecules of $H_2O$ and $CO_2$ colliding with molecules of nitrogen and oxygen -- a minor process within a gas, but nevertheless present.)

Some preliminary background on the thermodynamics of the atmosphere is provided here prior to the discussion of the quantification of those three processes. The first law of thermodynamics is a statement of the conservation of energy for a thermodynamic system:

$$dH = du + dW$$

where $dH$ = an infinitesimal amount of heat added per unit mass; $du$ = change in internal energy per unit mass; and $dW$ = work done by unit mass of the system.

In an *adiabatic process* here is *no heat exchange* between the system and the environment.

$dH = 0$ and the first law can be written as:

$$0 = C_p\, dT - \alpha\, dP \text{ or } C_p\, dT = \alpha\, dP$$

where $\mathbf{C_p}$ = **specific heat** at constant pressure = 1001.6 $[m^2\ s^{-2}\ K^{-1}]$ = 0.2394 [cal /g K]; T = temperature in degrees Kelvin (K), $\alpha$ = specific volume = 1 / density ($\rho$), P = pressure = $\rho\, R\, T$ (in an ideal gas), and R is the gas constant for dry air. The change of temperature with height is given by the "lapse rate" = $-\,\partial T/\,\partial Z$.

The lapse rate for an *adiabatic process* can be provided as follows. Neglecting vertical acceleration and friction in the vertical equation of motion (which are relatively small) for the atmosphere, the hydrostatic equation results: $dP = - g \rho \, dZ$

which simply states what most people already know – the change in pressure decreases as the height above the ground (Z) increases. Using this in the *adiabatic process:*

$C_p \, dT = \alpha \, dP$; but $\alpha = 1/\rho$ and the hydrostatic equation gives $dP = - g \rho \, dZ$, thus

$C_p \, dT = (1/\rho) (- g \rho \, dZ)$ g is the acceleration due to gravity $[m \, s^{-2}]$

$C_p \, dT = - g \, dZ$

$- \partial T / \partial Z = g / C_p = 9.8 / 1001.6 \, [m \, s^{-2}] / [m^2 \, s^{-2} \, K^{-1}] = \sim 9.8 \times 10^{-3}$ K/m $= \sim 9.8$ K / km. This adiabatic lapse rate *does not apply to the real atmosphere* and the three processes that are diabatic.

The three processes are tied to the lapse rate of the Standard Atmosphere, which is used internationally for many purposes – including aviation which is 6.5 K / km.

Sorokhtin, et al. 2007[2] have provided a way to quantify these processes in relative terms. Their procedure is clever and provides representative values for the three forces – at least in a space/time averaged sense. The detailed calculations are provided in Appendix C and summarized below where $C_w$ and $C_r$ are components of specific heat for water vapor and radiation, respectively.

$(C_p + C_w + C_r) = R / \mu \alpha = (1.987) / (29) (0.1905) = 0.3597$

$C_p = 0.2394 \, [cal / g \, °K]$

$C_w = (R / \mu \alpha) (T_E / T_S) - C_p = (0.3597) (263.6 / 288.2) - 0.2394$
$\quad = 0.0896 \, [cal / g \, °K]$

$C_r = (R / \mu \alpha) (T_S - T_E) / T_S = (0.3597) (24.6 / 288.2)$
$\quad = 0.0307 \, [cal / g \, °K]$

The sum of the three forces is $(R / \mu \, \alpha) = 0.3597$, thus the relative roles of each are:

Diabatic convection: $0.2394 / 0.3597 = 66.56\%$

Diabatic condensation of water vapor
$= 0.0896 / 0.3597 = 24.91\%$

Diabatic radiation (primarily $H_2O$ and $CO_2$)
$= 0.0307 / 0.3597 = 8.53\%$

These values are useful, but are at best ensemble averages over space/time -- there can be significant variability among the three processes as assured by the chaotic nature of the atmosphere. There is a large range of convection even with no visible storms present. Mix in summer and winter storms and values of intensities of the three forces will vary on any given day.

The interaction of the wind and mass field contribute to the diversity of the atmosphere. However, the *physics of radiation is God given* – the $CO_2$ molecules do not change, their surface absorption coefficients are a *function of wavelength dictated by their molecular structure.* The Planck function which establishes the intensity of radiation is a function of wavelength and temperature – and has been quantified by Max Planck – *it is not a function of chaos.* The *radiation force is special* -- though with the least magnitude, it is perhaps the most important force.

Convection at a given point has a very wide range. Latent heat release varies from zero to large amounts. Radiation is working everywhere all the time – it is precise and operates near the speed of light – acting together with the other two forces as an integrator to achieve energy balance.


## Earth's Energy Budget

Perhaps one may better understand the radiative roles of water vapor and $CO_2$ in the creation of the low level thermal blanket, the subsequent transfer of energy to higher levels, and the final achievement of energy balance, if the larger picture of Earth's radiative budget is provided.

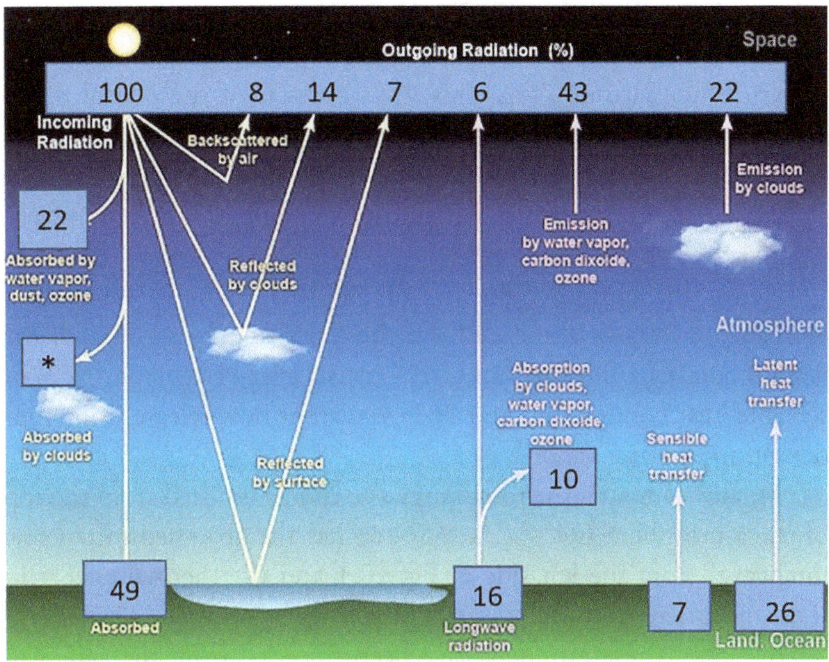


Figure 10.1  Energy budget from Stephens, et al with a 29% albedo

The diagram in Fig 10.1 is from Peixoto and Oort[3], but the numbers in the boxes are *slightly revised* for the energy balance for the decade 2000-2010 from Stephens, et al.[4]. The mean incident solar radiation received by the Earth is 340 Watts m$^{-2}$, considered as 100%; all other values are expressed in % terms. The first measured value of the Earth's albedo by Vonder Haar and Suomi[5] (the percent of solar energy reflected back to space) was 30%, the value used by the Stephens team was 29%.

The distribution of the incoming 100% of solar radiation is as follows: *29%* is reflected (backscatter by air 8%, reflected by clouds 14% and reflected by the Earth surface 7%).

On the far left of Fig 10.1 is the account of the *71%* of solar energy that was not reflected (100% minus the albedo of 29%): *22% of solar absorbed* by the atmosphere [water vapor, dust, ozone and clouds – with the numerical value for clouds not provided (*)], and the balance of *49%* absorbed by the Earth's surface.

Across the top right half of Fig 10.1 is the matching longwave radiation to space of *71%.* This *71%* consists of the *6%* of direct surface-to-space emission through the $H_2O$ and $CO_2$ window regions (discussed in the next Section), *43%* of atmospheric radiation loss via $H_2O$, $CO_2$ and ozone from the atmosphere, and *22%* emission from clouds within the atmosphere.

Across the bottom right half of Fig 10.1 is the required *49%* emission of longwave energy from the Earth's surface to match the 49% incoming from the Sun (*16%* via *net* surface radiation, *7%* of sensible heat conduction, and 26% via latent heat release from condensation of water vapor).

Finally, to match the total longwave energy emitted from the top of the atmosphere *(43 + 22 = 65)* one has the absorbed solar (now longwave) from the left side of Fig 10.1 and the thermal energy in the atmosphere from the lower right giving the total of *(22 + 10 + 7 + 26 = 65)*. These numbers have been rounded from the values in Stephens[4] which have ± values reflecting uncertainty in his measurements. These values are similar to those from other references.

There remains a trivial detail to consider. The *solar energy* in the left side box of Fig 10.1 indicates solar energy absorbed directly by "water vapor, dust and ozone". There is also small amount of solar absorption by $CO_2$ in the atmosphere *above the surface*. This is a very small amount and the absorption coefficients are small at these short wavelengths. Moreover, the *Planck radiation intensity* is *very small* because of the short wavelengths and lower temperatures. This trivial delta-heat radiates to space as that of $H_2O$ and $CO_2$. [This may be clearer after reading Chapter 11].

## Entropy Change

The concept of entropy is quite important and perhaps should be explained in some detail. Entropy is a measure of disorder and in the universe there is far more disorder than order. Entropy increases over time. Buildings fall into disrepair if left unattended. Ancient structures fall into ruin and crumble over time. Automobiles rust, dramatic rock formations eventually erode, and people age. There is no escaping the second law of thermodynamics – everything decays – disorder always increases.

The universe naturally evolves toward disorder, one must *expend energy* to create order or structure. In order for energy to perform work, a difference must exist between energy at a high potential and energy at a more randomized, diluted potential.

The term *entropy* is a measure of the degree to which energy has lost the capacity to do useful work. The change of entropy (S) over time of a system is given by $dS/dt \geq Q/T$ where Q is the diabatic heating transferred at temperature T. Irreducible, diabatic processes create entropy production [$dS/dt > 0$].

The Earth receives high-quality "rich" energy from the Sun (low entropy) and returns low-quality "impoverished" energy to space (high entropy). The estimated entropy change is shown in Table 10.1 from energy data from Stephens[4] and the temperatures are from Peixoto and Oort[3].

The change in entropy due to the incoming solar radiation and the outgoing longwave terrestrial radiation is provided in Table 10.1. The entropy is given by the diabatic heating Q which is obtained from Figure 10.1 (where E = Q) with the incoming radiation as 340 W m$^{-2}$

From Table 10.1, the first line in each column is the active temperature, the second line is the percent of 340 for each category (e.g., the incoming solar energy is 340 x 0.71 since the albedo is 29%), the third line is the resultant E, the fourth and fifth lines give the value of the entropy in milli-Watts m$^{-2}$ per degree K. Column one indicates the incoming entropy as 41.9 m-W m$^{-2}$ K$^{-1}$.

Table 10.1  Average annual entropy incoming and leaving the
Earth during 2000-2010

| Solar incoming | Longwave From surface | Longwave From Clouds | Longwave From Atmosphere |
|---|---|---|---|
| T = 5760 K | T = 288 K | T = 259 K | T = 252 K |
| E or Q = 340 x 0.71 | E or Q = 340 x 0.06 | E or Q = 340 x 0.22 | E or Q = 340 x 0.43 |
| E = 241.4 W m$^{-2}$ | E = 20.4 W m$^{-2}$ | E = 74.8 W m$^{-2}$ | E = 146.2 W m$^{-2}$ |
| S = E / T = 0.0419 | S = E / T = - 0.0708 | S = E / T = - 0.2888 | S = E / T = - 0.5802 |
| = 41.9 milli-W m$^{-2}$/K | = - 71 m-W m$^{-2}$ / K | = - 289 m-W m$^{-2}$ / K | = - 580 m-W m$^{-2}$ / K |

The 2nd column gives the longwave radiation directly from the surface to space via the $H_2O$ and $CO_2$ windows. This is 6% of 340 or 20.4 W m$^{-2}$. From the surface temperature of 288 K, this indicates the amount of entropy of -71 m-W m$^2$ K$^{-1}$ [negative because leaving the Earth system.]

The 3rd column indicates the longwave energy from clouds to space of 74.8 W m$^{-2}$. The active temperature from the reference[1] is 259 K which gives the entropy = -289 m-W m$^{-2}$ K$^{-1}$. A similar calculation from the information in the 4th column gives: - 580 m-W m$^{-2}$ K$^{-1}$.

The total outgoing entropy is 940 m-W m$^{-2}$ K$^{-1}$; the incoming was 41.9 m-W m$^{-2}$ K$^{-1}$. *The gain in entropy is over a factor of 22.43 ~ =22. For each photon of solar energy received, there are 22 photons of longwave energy (heat) sent to space. The original data from Peixoto and Oort had similar numbers of 925 of outgoing and 41.3 of incoming for the ratio of 22.40 ~ = 22.* A solar physicist[6] performed a similar estimate in 1982 for the entropy increase for the outgoing atmosphere entropy versus the incoming solar entropy and came up with a *gain factor of 17.*

The final transmission of sufficient heat to space for energy balance is provided by *radiation* – the previous calculation had the relative magnitude of $H_2O$ and $CO_2$ radiation as *minor* -- compared to the role of latent heat and convection. However, radiation, is working everywhere all the time (24 hours a day) acting as an integrator for the three forces to achieve energy balance.

The total energy of the universe is constant and the entropy, the non-useable energy, is constantly increasing. This is the second law of thermodynamics – well studied by Planck and many others. This law outlines the transfer of heat from warm bodies to cold. Heat and work are entities by which systems exchange energy with one another. Heat dissipation makes work and life possible.

In Chapter 2 the discussion of the formation of the universe was presented. The eventual concern of whether the universe would expand, contract or remain basically the same was eventually resolved. The answer was that the universe was expanding – even more important, the expansion was accelerating!

In 2010 there was an estimate of the entropy of the universe made by Egan and Lineweaver[7]. The greatest source of increased entropy is from supermassive black holes and the estimate of the entropy of the observable universe is $3.1 \times 10^{104}$ k where k is Boltzmann's constant: k = $1.38 \times 10^{-23}$ Joules $K^{-1}$.

The universe is expanding, and accelerating -- the above authors have gone on and computed the entropy of our universe which is now entering dark energy domination with an equation of state of w > -1 (radiation and matter).

The entropy of the cosmic event horizon (CEH) has slowed and has the value of $2.6 \pm 0.3 \times 10^{122}$ k and is almost as large as it will ever become. During dark energy domination, the proper distance to the CEH is time dependent and the proper radius, volume and entropy of the CEH monotonically increase to constant values. When the universe is *dominated* by dark energy w = -1 and the CEH entropy is $2.88 \pm 0.16 \times 10^{122}$ k.

## References: Chapter 10


1.  Emanuel, K.A., 1994: *Atmospheric convection*. Oxford University Press, New York, 580 pp.

2.  Sorokhtin, O. G., G. V. Chilingar and L. F. Khilyuk, 2007: *Global warming and global cooling: evolution of climate on Earth*. Elsevier, Amsterdam, 313 pp.

3.  Peixoto, J. P. and A. H. Oort, 1992: *Physics of climate*. American institute of Physics, New York, 522 pp.

4.  Stephens, G. L., J. Li, M. Wild, C. A. Clayson, N. Loeb, S. Kato, T. L'Ecuyer, P. W. Stackhouse Jr, M. Lebsock and T. Andrews, 2012: An update on Earth's energy balance in light of the latest global observations. *Nature Geoscience*, **5**, 691-696.

5.  Vonder Haar, T. H. and V. E. Suomi, 1969: Satellite observations of the Erath's radiation budget. *Science*, **163**, 667-669.

6.  Frautschi, S., 1982: Entropy in an expanding universe. *Science*, **217**, 593-599

7.  Egan, C. A. and C. H. Lineweaver, 2010: A larger estimate of the entropy of the universe. arXiv: .3983v309. [Astro - ph.CO] 25 Jan 2010

Liou[4] defines an optical depth Tau = $\mathcal{T}$ = $\int K_\lambda(z)\,\rho(z)\,dz$ so that $d(\mathcal{T})$ = $K_\lambda(z)\,\rho(z)\,dz$

Liou further defines the monochromatic transmittance such that the exponential attenuation of radiation can be expressed as $T_\lambda(\mathcal{T}) = e^{-\mathcal{T}}$ and the differential form equals $dT(\mathcal{T})\,/\,d\mathcal{T} = -1\,e^{-\mathcal{T}}$. The algorithm for *Solution #1* below is derived in Appendix D from equations in Liou[4].

Liou's integral equations lead directly to the *upward flux* (F) of radiance given by the following:

$$F(I) = e^{-\Delta\mathcal{T}}\,F(I-1) + [1 - e^{-\Delta\mathcal{T}}]\,B(\lambda,T) \text{ where } \Delta\mathcal{T} = \boldsymbol{\Delta\ tau}$$
$$= \boldsymbol{D\ K_\lambda(z)\ \rho(z)\ dz\ \textit{(where D is a constant)}}$$

The proper units for $\Delta$ tau must be $[m^2/kg]\,[kg/m^3]\,[\mathbf{m}]$ = $[m^3/kg]\,[kg/m^3]$. Thus dz must be in **meters**. However the $CO_2$ absorption coefficients in Table 11.1 were expressed as changes over a *kilometer*, so those absorption coefficients shown above must be divided by 1000 – with 596.1 becoming 0.5961. [In the Appendix D for Solution #2, the coefficients will remain as originally presented with the largest coefficient in the 15 µm region being 596.1 $m^2$ / kg as in Table 11.1].

The integration of the *Schwarzschild equation* is required. The largest impact from level to level in the atmosphere is the change in the intensity of the *Planck equation* (illustrated in Fig 11.2). Note how it changes with temperature and wavelength). The formula is:

$B(\lambda, T) = [2\,h\,c^2\,/\,\lambda^5]\,[\exp(ch\,/k\,\lambda\,T) - 1]^{-1}$ where h = Planck's constant = $6.63 \times 10^{-34}$ J s, k is the

Boltzmann's constant = $1.38 \times 10^{-23}$ J / K, and
$$c = \text{velocity of light} \sim = 3 \times 10^8 \text{ m/s}$$

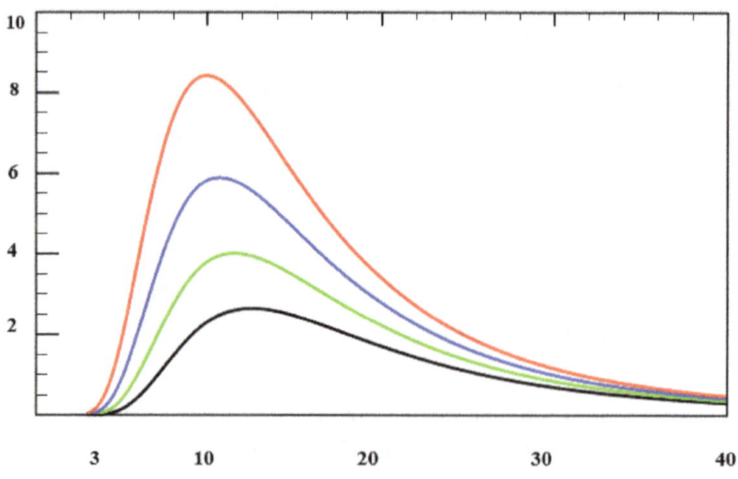

**Planck Intensity versus wavelength (microns) for different temperatures (K)**

230  Black      250 Green      270 Blue      290 Red

Fig 11.2  Radiative power intensity expressed by Planck's equation W m$^{-2}$ ster$^{-1}$ μm$^{-1}$

Consider the transfer of thermal infrared radiation emitted from the Earth and the atmosphere where a beam of intensity will undergo the *absorption* and *emission* processes simultaneously. The *Schwarzschild equation* for this process is: $dI_\lambda / k_\lambda \rho \, ds = - I_\lambda + B_\lambda (T)$; the first term on the right hand side denotes the reduction of radiant intensity due to absorption of radiation from the bottom of the layer to the top, whereas the second term represents the increase in radiant intensity arising from the blackbody emission of the material within the layer.

The atmosphere is considered to be in thermodynamic equilibrium and plane parallel. A *differential optical depth can be defined as $d\tau = - k_\lambda (z) \rho (z) \, dz$. The coordinates below from Liou.*

$$0 \; Z_\infty \; \text{---} \; \text{---} \; \text{---} \; \text{---} \; \text{---} \; \text{---} \; \text{---} \; \text{---} \; \text{---} \; \text{---} \; \text{---} \; \text{---} \; \text{---} \; \text{--} \; T_\infty \; u_1 \; 0$$

$$\tau' \; Z' \; \text{---} \; \text{---} \; \text{---} \; \text{---} \; \text{---} \; \text{---} \; \text{---} \; \text{---} \; \text{---} \; \text{---} \; \text{---} \; \text{---} \; \text{---} \; \text{---} \; T' \; u' \; P'$$

$$\tau \; Z \; \text{-------------------------------------------------} \; T \; u \; P$$

$$\tau' \; Z' \; \text{---} \; \text{---} \; \text{---} \; \text{---} \; \text{---} \; \text{---} \; \text{---} \; \text{---} \; \text{---} \; \text{---} \; \text{---} \; \text{---} \; \text{---} \; \text{---} \; T' \; u' \; P'$$

$$\tau_s \; 0 \; \underline{\hspace{10cm}} \; T_S \; 0 \; P_S$$

The coordinate systems in $\tau$, **Z**, **u**, **T** and **P** for IR radiative transfer are shown above. *The path length (u)* is for absorbing and emitting gases (they absorb and warm and emit and cool) defined for the surface upward. The total path length is defined as $u_1$. $T_\infty$ and $Z_\infty$ are temperature and height at the top of the atmosphere. The surface temperature = $T_S$. The surface pressure is $P_S$. *Z is a reference level. The optical thickness of the ith layer is $\Delta \, tau = D\rho_i \, K_\lambda \, \Delta z_i$.*

The Schwarzschild integrations from Liou are then provided below (with the diffusion term **D = 5/3** added for diffuse radiation). Note that $e^{-D} = 1 / e^{5/3} = 1 / 5.29445 = 0.1888756$ and $1 - e^{-D} = 0.811124$. The radiation Flux (F) intensity is $\Pi \times B(\lambda,T)$ to put in units of watts/m², but to simplify the notation the *Pi* term and its effects will be shown later; see footnote 7 at the end of this Chapter.

*The algorithm for Solution #1 is derived from Liou's integrals in Appendix D. The net radiation* or *back radiation* for a given level (that sent upward at the bottom of a layer, minus that sent downward at the top of a layer) must also be computed.


## Divide the atmosphere into N layers from 0 to N.

*For upwelling radiance*

Begin with the blackbody radiation Flux at the surface:
$$F(0) = B(T_0)$$

Iterate *upward* with I increasing {I = 1 to 18 as an example}

$F(I) = [e^{-\Delta \tau}] F(I-1) + [1 - e^{-\Delta \tau}] B(I)$    i.e.
$F(1) = [e^{-\Delta \tau}] F(0) + [1 - e^{-\Delta \tau}] B(1)$

*For down welling radiance*

Start at top of the atmosphere where F(I) = 0, use {F(19) = 0 as an example}

Iterate *downward* with I decreasing    {I = 18, 1, -1 as a FORTRAN example}

$F(I) = [e^{-\Delta \tau}] F(I+1) + [1 - e^{-\Delta \tau}] B(I)$    i.e.
$F(18) = [e^{-\Delta \tau}] F(19) + [1 - e^{-\Delta \tau}] B(18)$

One should test the integrity of this algorithm. Suppose $CO_2$ did not absorb radiation – all the absorption coefficients are zero: thus all $K_\lambda = 0$ and therefore $e^{-\Delta \tau} = 1$ and $[1 - e^{-\Delta \tau}] = 1-1 = 0$.

Start with the surface $B(T_0) = B(0)$ for the *upwelling radiance* one would have [with F(0) = B(0) as required]:

$F(I) = [e^{-\Delta \tau}] F(I-1) + [1 - e^{-\Delta \tau}] B(I)$
$F(1) = (1) F(0) + (0) B(1) = (1) B(0) + 0 = B(0)$
$F(2) = (1) F(1) + (0) B(2) = (1) B(0) = B(0)$
$F(N) = (1) F(N-1) + (0) B(N) = (1) B(0) = B(0)$

Thus, the intensity of the radiation is the same all the way to the top as it should be since there is no absorption.

The algorithm for the *down welling radiance* provides the following [starting with F(19) = 0]:

$$F(I) = [e^{-\Delta\tau}] \, F(I+1) + [1 - e^{-\Delta\tau}] \, B(I)$$

$$F(18) = (1) \, F(19) + (0) \, B(18) = (1) \, (0) + (0) = 0$$

$$F(17) = (1) \, F(18) + (0) \, B(17) = (1) \, (0) + (0) = 0$$

$$F(1) = (1) \, F(2) + (0) \, B(1) = (1) \, (0) + (0) = 0$$

Thus, the down welling radiation is zero at all levels, and the *net radiation* for each level is that which went up minus zero equals the original unchanged intensity. *The algorithm is consistent.* It works well in nearly all cases, but there are two flaws which will be discussed. This is why a slightly better method with neither flaw has been created and is outlined in the Appendix D.

Table 11.3  Planck intensity: $B(\lambda, T)$ for the strongest coefficient in Band 1 and in Band 2

| Height (km) | Temperature | $B(\lambda,T)/B(\lambda,T_0)$ $\lambda = 4.23466$ | $B(\lambda,T)/B(\lambda,T_0)$ $\lambda = 14.9815$ | $CO_2$ density $\rho(T)/\rho(T_0)$ | $T/T_0$ |
|---|---|---|---|---|---|
| 0 | 288.15 | 1.0000 | 1.0000 | 1.0000 | 1.0000 |
| 1 | 281.65 | 0.7619 | 0.9235 | 0.9075 | 0.9774 |
| 3 | 268.65 | 0.4252 | 0.7791 | 0.7422 | 0.9323 |
| 5 | 255.76 | 0.2236 | 0.6466 | 0.6010 | 0.8872 |
| 7 | 242.65 | 0.1098 | 0.5265 | 0.4813 | 0.8421 |
| 9 | 229.65 | 0.0497 | 0.4192 | 0.3807 | 0.7970 |
| 11 | 216.15 | 0.0205 | 0.3251 | 0.2971 | 0.7519 |
| 13 | 203.65 | 0.0075 | 0.2443 | 0.2283 | 0.7067 |
| 15 | 190.65 | 0.0024 | 0.1768 | 0.1725 | 0.6616 |
| 17 | 177.65 | 0.0007 | 0.1220 | 0.1277 | 0.6165 |

The Planck *differential with height is very large* for the short wavelengths in Band 1 – as indicated in Table 11.3 and evidenced in Fig 11.2. The Planck intensity is *virtually transparent already at 11 km using Solution #1 or Solution #2* as shown in Appendix D.

Runs over Band 1 where the largest absorption coefficient is 4596 $m^2$/kg are found in Appendix D. The runs in this Chapter will only be over the wavelengths of Band 2 where the important absorption is near the 15µ region. Band 2 is far more interesting and a *represents a much stronger challenge to prove that $CO_2$ has no impact on climate change.*

A run over Band 2 with *Solution #1* is provided in Table 11.4. This run includes 25,001 lines over the wavelength region of 7.98133 to 17.98133 µm. This run covers the important Band with the large absorption coefficients near 15 µm of $K_\lambda$ *= 596.1 $m^2$ /kg* at 14.98133 µm. *Solution #1* requires $K_\lambda$ be divided by 1000 as discussed above: thus $K_\lambda$ *= 0.5961.*

The run shown below in Table 11.4 has only 25,001 lines, however, Appendix D shows the same run indicated in Table 11.4, but with two runs (one of 25,001 lines the other with 50,001 lines – both of these runs used *Solution #2* with the maximum coefficient of 596.1 *Both of those runs produced the same answer as the data sets are statistically equal.]*

Recall that $H_2O$ and $CO_2$ do not create energy – both merely distribute that energy upward – along with the other two more powerful processes -- latent heat release and convection.

Table 11.4  Solution #1 over Band 2 with 25,001 over 7.98133 to 17.98133 µm.

| Line Number (J) | 11501 | 13501 | 15501 | 17501 | 19501 |
|---|---|---|---|---|---|
| Wavelength Microns | 12.5813 | 13.3813 | 14.1813 | 14.9813 | 15.7813 |
| Absorption Coefficient | $0.742 \times 10^{-6}$ | $0.965 \times 10^{-5}$ | $0.644 \times 10^{-3}$ | 0.5961 | $0.679 \times 10^{-3}$ |
| Planck value Surface | 7.29578 | 6.82998 | 6.34141 | 5.85274 | 5.37878 |
| Planck value 16 km | 7.29570 | 6.82903 | 6.28344 | 0.20716 | 5.32852 |
| Planck value 16 km / Sfc. | 100% | 99.99% | 99.09% | **3.54%** | 99.07% |

The line numbers in Table 11.4 are separated deliberately by 2000 lines which were selected from the 25,001 lines of the calculation. This produced a *random selection* of the other absorption coefficients – apart from the deliberate attempt to view the *maximum coefficient* at line number 17501 with the maximum absorption coefficient of *0.5961.*

It is seen that these other surface coefficients ranged from 0.742 x $10^{-6}$ to 0.679 x$10^{-3}$. These are so incredibly small that there is virtually no heat absorption by these $CO_2$ coefficients – the ratio of Planck intensity [16 km / Sfc.] is ~100%. On the other hand, the maximum absorption coefficient has been effective in reducing the original Planck intensity by the factor of > 28 at 16 km. Appendix D indicates that 86.34 % of the surface coefficients in this range are transparent.

Some scientists solving the Schwarzschild equations using the algorithm above for *Solution #1* make two assumption for the N layers in the integration. One assumption is that the layers are isothermal (so there is no temperature effect on the absorption coefficient – though in theory the coefficient does become very slightly stronger with decreasing temperature). This is not a bad assumption if the layers are thin. There is another assumption, perhaps used by some, that is bad.

Table 11.5  Net Flux with height for *Solution #1* for Methods 1 through 3

| Height of Layer in km | | Method 1 | Method 2 | Method 3 | Method 4 |
|---|---|---|---|---|---|
| 1 | Flux Up | 5.482 | 5.555 | 5.568 | 5.376 |
| | Flux Down | 5.291 | 5.118 | 5.151 | 4.938 |
| | Flux Net | 0.191 | 0.437 | 0.417 | 0.438 |
| 2 | Flux UP | 5.076 | 5.168 | 5.190 | 4.912 |
| | Flux Down | 4.839 | 4.698 | 4.731 | 4.486 |
| | Flux Net | 0.237 | 0.470 | 0.459 | 0.426 |
| 4 | Flux Up | 4.303 | 4.362 | 4.392 | 4.029 |
| | Flux Down | 3.981 | 3.910 | 3.943 | 3.632 |
| | Flux Net | 0.322 | 0.452 | 0.449 | 0.397 |
| 6 | Flux Up | 3.605 | 3.606 | 3.637 | 3.215 |
| | Flux Down | 3.185 | 3.194 | 3.226 | 2.854 |
| | Flux Net | 0.420 | 0.412 | 0.411 | 0.361 |
| 8 | Flux Up | 2.985 | 2.921 | 2.952 | 2.484 |
| | Flux Down | 2.453 | 2.553 | 2.585 | 2.165 |
| | Flux Net | 0.532 | 0.368 | 0.367 | 0.319 |
| 10 | Flux Up | 2.445 | 2.311 | 2.343 | 1.847 |
| | Flux Down | 1.793 | 1.989 | 2.020 | 1.574 |
| | Flux Net | 0.652 | 0.322 | 0.323 | 0.273 |
| 12 | Flux Up | 1.983 | 1.779 | 1.810 | 1.310 |
| | Flux Down | 1.215 | 1.503 | 1.532 | 1.087 |
| | Flux Net | 0.768 | 0.276 | 0.278 | 0.223 |
| 14 | Flux Up | 1.600 | 1.325 | 1.354 | 0.878 |
| | Flux Down | 0.737 | 1.090 | 1.119 | 0.705 |
| | Flux Net | 0.863 | 0.235 | 0.235 | 0.173 |
| 16 | Flux Up | 1.289 | 0.948 | 0.976 | 0.549 |
| | Flux Down | 0.365 | 0.732 | 0.768 | 0.423 |
| | Flux Net | 0.924 | 0.216 | 0.208 | 0.126 |

Method 1 The density in each layer is the *average* for the top and bottom of each layer

Method 2 Density in each layer is $\rho(I-1)/\rho(I)$ going UP and opposite coming Down

Method 3 Density is nonlinear $B(I)/B(I-1) \times \rho(I-1)/\rho(I)$ going UP; opposite coming Down

Method 4 Density is nonlinear as above; but using *Solution #2* in Appendix D. [The Ratio of Planck intensity (16 km / Surface) for Method 3 is 0.208/5.853 = 3.6% and the ratio for Method 4 is 2.16% -- *(both Solutions #1 and #2 are within 2 % of each other for this strong coefficient)*.

The second assumption sometimes made (and it is not possible to estimate how many scientists or climate modelers do this) is to use the *average CO$_2$ density in a layer.* This is *not* a good assumption with *Solution #1. This is shown by the various methods in Table 11.5!*

The density of CO$_2$ = [P / 1.889 T]. The density decreases with height as pressure (P) decreases faster with height than *temperature* (T). The decrease in density would *increase* the relative radiation intensity. Note, however, that the Planck function is a *strong function of both wavelength and temperature. The use of a mean density leads to a significant positive bias of intensity in the upper atmosphere. The proof of this bias is illustrated in Table 11.5.*

The Planck function has a stronger impact with temperature *decreasing continuously* through the layer and a better approach may be to add the Planck effect with density in a *nonlinear way* -- compensate the *intensity increase with decreasing density*, by the *decrease* in Planck intensity due to the *temperature decrease with height.* Such a change would be added to the tau term, such as:

*[B (I) / B (I-1)] x ρ (I-1) / ρ (I) or as an example:*
*[B (1) / B (0)] x [ρ (0) / ρ (1)].*

Table 11.5 indicates the *positive intensity bias in the upper atmosphere* comes from Method 1 which uses an *average density* in each layer -- which is the same whether going up or down. Method 1 intensity is *stronger* than the other Methods going up and is *weaker* than the other Methods coming down, so the Net *increases* with height (this is not reality, the Net should *decrease* with height as the other Methods show) and has too much intensity in the upper atmosphere.

If all modelers used this *average density*, it would partially account for the extra heat observed in their upper atmosphere.

Methods 2, 3 and 4 give approximately the same answers. Method 3 has the nonlinear change for $\Delta$ $CO_2$ density for *Solution #1* and Method 4 has the same for *Solution #2*. These are within 2% of each other.

For the purposes of this study, the impact of the $CO_2$ coefficients over the full range of wavelengths should be performed. This has been accomplished with many runs of different lengths over a variety of wavelengths. Further examples are shown or discussed in Appendix D.

The main point that results from all *these different calculations* is that the *largest coefficient commands* the most attention and using *just this maximum coefficient* can simplify some comparisons. This single line approach will be evaluated below with various *different temperature profiles* to see if significant differences result.

The temperature profile used for all calculations shown so far was that for the Standard Atmosphere. However, a very wide range of temperature profiles scattered about that Standard (some stable, some unstable) were evaluated – and a few of these are shown in the Table 11.6

All these runs have the *nonlinear delta $CO_2$ change* and all have used the delta temperature change over a layer [which is T(I) / T(I-1)] going up and reversed going down]. Columns 2 and 3 in Table 11.6 look at temperatures colder and warmer than the Standard in the *lower levels below 5 km.*

The other lapse rate profiles in the last two columns on the right of the Table show that a *colder upper atmosphere* (> 5 km) leads to a slightly greater reduction (3.21 %) but the warmer upper atmosphere gave a higher answer than the Standard Atmosphere Lapse Rate.

Note that the colder upper atmosphere results in even less intensity than that of the Standard Temperature Lapse Rate. But slightly greater intensity for the warmer upper atmosphere case.

Table 11.6 Net Planck Intensity for various temperature profiles $K_\lambda = .5961$ from *Solution #1*

| Net Planck Intensity at Level (km) | Standard Atmospheric Profile | Colder lower Entire profile -5 *Surface to 5 km* | Warmer lower Entire profile +5 *Surface to 5 km* | Colder upper air Entire profile − 5 *From > 5 km to 19 km* | Warm upper air Entire profile +5 *From > 5 km to 19 km* |
|---|---|---|---|---|---|
| Surface | **5.85274** | **5.85274** | **5.85274** | **5.85274** | **5.85274** |
| 2 | 0.4566 | 0.4302 | 0.4803 | 0.4594 | 0.4524 |
| 4 | 0.4491 | 0.3473 | 0.5312 | 0.4700 | 0.4181 |
| 6 | 0.4113 | 03641 | 0.4688 | 0.5391 | 0.3036 |
| 8 | 0.3681 | 0.3614 | 0.3763 | 0.3720 | 0.3674 |
| 10 | 0.3235 | 0.3226 | 0.3247 | 0.3095 | 0.3382 |
| 12 | 0.2788 | 0.2786 | 0.2789 | 0.2621 | 0.2956 |
| 14 | 0.2359 | 0.2359 | 0.2360 | 0.2190 | 0.2532 |
| 16 | **0.207155** 3.54% | **0.207152** 3.54% | **0.207159** 3.54% | **0.1880** 3.21% | **0.2272** 3.88% |

The last row in the Table 11.6 indicates that the ratio of the Planck intensity at 16 km versus the Surface value is the same in these two columns -- whether the lower layer is cold or warm. However the first 6 km do reflect the decrease (cooler) or increase (warmer) in net Planck intensity, then catch up and match the Standard Atmosphere Net at 16 km. It is seen that the value for the Standard Atmosphere Profile is 3.54% at 16 km. It has fallen by a factor > 28 times.

The above profile changes are not very significant. Nevertheless it may be useful to compare the results from *Solution 2* (discussed in detail in Appendix D) and shown in Table 11.7. These are the same temperature profiles shown in Table 11.6.

The difference between the two *Solution #1* and *Solution #2* is indicated as follows from results in Tables 11.6 and 11.7. The Standard Atmosphere Lapse Rate difference is (3.54 % -2.16 % = 1.38 %). The **Colde**r upper atmosphere difference is (3.21 % - 1.64 % = 1.57 %). The **Warm** upper air difference is (3.88 % - 2.78 % = 1.10 %). *Thus the differences in all case were less than 2 %.*

The concentration of $H_2O$ at the one km level alone is capable of absorbing all the available solar heat at the surface, and does absorb 5 times that of $CO_2$. *All the Sun's heat available at the surface is fully re-distributed vertically by all the molecules with the help of all the coefficients.*

The solar input at the surface varies with cloud cover, and of course with the four seasons of the year as the Earth traverses its path about the Sun. The temperature of the thermal blanket varies accordingly. *Adding passing clouds in some statistical way would lower the above numbers slightly (as studies show that clouds have a net cooling effect),* thus a dry atmosphere provides a more stringent proof.

There is also the small, perhaps insignificant effect of molecular collisions of $CO_2$ molecules with nitrogen and oxygen which would also lower the $CO_2$ radiative intensity to a slightly lower level.

One can summarize these calculations as follows: whatever the *climate-change regime, whatever surface heat from the Sun is available on any given day – based upon the weather variability of that day--*

*within that climate-change regime, that heat is fully absorbed and fully vertically redistributed throughout the troposphere – $CO_2$ both absorbs and emits radiant heat in a systematic way – no net climate-change is produced.*

Table 11.7 Net Planck Intensity for various temperature Profiles
$K_\lambda = 596.1$ from *Solution #2*

| Net Planck Intensity at Level (km) | Standard Atmospheric Profile | $K_E$ defined in Appendix C For Standard (Solution #2) | Colder upper air Entire profile – 5 From > 5km to 19 km | Warm upper air Entire profile +5 From > 5 km to 19 km |
|---|---|---|---|---|
| Surface | 5.85274 | 596.1 | 5.85274 | 5.85274 |
| 2 | 0.4260 | 3.638 | 0.4260 | 0.4260 |
| 4 | 0.3971 | 4.242 | 0.3971 | 0.3971 |
| 6 | 0.3613 | 3.623 | 3.660 | 0.4075 |
| 8 | 0.3194 | 2.673 | 0.2754 | 0.3670 |
| 10 | 0.2728 | 1.787 | 0.2298 | 0.3202 |
| 12 | 0.2232 | 1.100 | 0.1830 | 0.2688 |
| 14 | 0.1734 | 0.6247 | 01376 | 0.2152 |
| 16 | 0.1262<br>2.16 % | 0.3257<br>$K_E < 0.33$ | 0.0962<br>1.64 % | 0.1625<br>2.78 % |

Why does the integrated effect of $CO_2$ have so little effect on the total temperature profile? The *primary* reason is that the Planck function change with height (temperature) is *very strong* in reducing the intensity of those *relatively few lines with large absorption coefficients.*

A second reason is that terrestrial radiation is not a pencil beam, but diffuse radiation propagating in all directions. The way that diffuse radiation paths are handled is discussed further in detail in

Appendix D. Both radiation experts Houghton[5] and Liou[4] agree on the diffuse factor used above and in the Appendix. Thus, *longwave radiation is diffuse* and the radiation intensity depletes more rapidly over distance than a pencil beam of non-diffuse radiation.

*The net radiation* or *back radiation* for a given level (that sent upward at the bottom of a layer, minus that sent downward at the top of a layer) must be computed. This also leads to a further *minor* depletion of intensity with height.

## Summary of this important Chapter

Though radiation has the smallest percentage impact of the three physical processes discussed in Chapter 10, it has the most important role of maintaining radiative balance and does this very effectively. Radition proceeds at the speed of light in a vacuum, and operates only slightly slower within the Earth's atmosphere.

The only way to really understand diffuse longwave Planck radiation within the atmopshere is to actually calculate the detailed integration of the Schwarzschild equation of radiative transfer with a full complement of $CO_2$ lines and absorption coefficients —a very large number of such lines should be used and each line evaluated individually rather than in an approximated "band model".

The $2^{nd}$ method of solving the Schwarzschild equations in Appendix D, based upon a slightly different equation from Houghton, used a *coefficient of reduced intensity (which can also be considered* 'cumulative absorptivity'). This was used as an *additional measure of the reduction* in $CO_2$ radiation intensity with height *over and above the actual measurement* of the Planck radiation intensity. That additional measure $K_E$ from many atmospheric runs produced a critical value of < 0.333. With this value of $K_E$ or less, that layer was considered transparent.

Both *Solution #1* and *Solution #2* gave answers within 2 % of each other for the resultant Planck intensity and the *same level of the atmosphere* where the maximum absorption coefficient had resulted in a *trivial value* for the intensity – virtually transparent, and all other smaller coefficients leading to total transparency.

The CO$_2$ coefficients from the Pacific Northwest National Laboratory (also available from the HITRAN date center) had > 98 % of the surface coefficients already transparent with lines between 1 and 30 µm. Both runs gave essentially the same results with 100% transparency at 16 km.

In an earlier paper[6] there were two runs of 150,000 and 300,000 lines over the 1 to 30 µm range. The answers were the same in both of the long runs – because the two data sets were statistically equal.

The maximum *coefficient of reduced intensity,* was 0.69 at 9 km in both runs referred to above. That calculation was performed here again with the effect of diffusion term being 0.811124 rather than 0.6 (from the Houghton equation where the effect of the diffusion term 5/3 implied a reciprocal value). This new diffusion term slowed the reduction of the Planck intensity with height.

The answer for that new calculation was 16 km (rather than the 9 km). The higher valued diffusion term, used in those long run solutions, and now used in *Solution 2* in the Appendix is 0.811124. This changed the critical value for the *coefficient of reduced intensity* to 0.33.

The two Solutions discussed above both provide the level where the remaining heat is approximately 2.16% or 46 times less at 16 km than at the surface[7]. This slight residual of heat available at that altitude is radiated to space as the coefficient is virtually transparent. All the remaining smaller coefficients are transparent – also allowing heat to radiate to space un-impeded.

The historical observational record and these calculations indicate that the CO$_2$ concentration had no impact on *climate-change*. CO$_2$ and H$_2$O distribute heat from the thermal blanket and allow upper level cooling for a zero *climate-change* net effect. Similar calculations with the H$_2$O bands in Table 11.1 provided transparencies at even lower levels than CO$_2$ as expected from the smaller absorption coefficients. The residual heat in the upper atmosphere simply follows the $2^{nd}$ law of thermodynamics and seeks the cold reservoir of outer space – increasing the entropy.

Three reasons were given why the integrated effect of $CO_2$ has so little effect on the temperature profile. The *primary* reason is that the Planck function change with height (temperature) is *very strong* in reducing the intensity of those relatively few lines with large absorption coefficients.

A second important reason is that the terrestrial radiation is not a pencil beam, but diffuse radiation propagating in all directions. In a plane parallel atmosphere the only concern is the propagation of diffuse radiation upward and downward. The *diffuse* radiation intensity depletes rapidly over vertical distance.

A third reason is that the *net radiation* for a given level must be computed (that sent upward at the bottom minus that sent downward at the top of a layer). This also *slightly depletes* the intensity with height. This is quite small for $CO_2$ as these calculations have shown, and there is perhaps a reason that can be traced back to two previous results that have been shown and are consistent.

First one must go back to the energy diagram of Fig10.1 in Chapter 10. The value of 16% is shown for the net surface longwave (not counting the sensible heat flux nor the latent heat flux). In the original paper by Stephens et al. (reference 4 in that Chapter) they had the net at the surface of 345 down and 398 up for a surface net of 53 watts/m². Their units for the incoming solar radiation was 340 watts/m² so their net surface up was 53/340 =15.6 or ~ 16 % -- which matches the value in Fig 10.1.

We also saw earlier that the broad brush calculation of the radiation absorption in the lowest one kilometer level was ~ a factor of 5 of water vapor over $CO_2$. One can easily imagine that of the measured net flux of back radiation at the surface of 16 % -- then this was ~ 2.65 % coming from $CO_2$ and 5 times that amount or ~ 13.25% from water vapor (add 0.1 % from low level clouds.)

One could add a fourth reason; all the CO$_2$ molecules are affected by all the absorption coefficients and 98 % of those CO$_2$ surface absorption coefficients from 1 to 30 μm had absorption coefficient values < 1.0 when those coefficients were for changes over a km. Using the *Solution #1* to solve the Schwarzschild equations requires that these be divided by 1000.

The equations of *Solution #1,* the standard solution for the integration of the Schwarzschild equations used by virtually all the modeling community *does an excellent job.* It has a minor flaw at extremely large values found in Band 1. This last point is demonstrated in Appendix D.

All longwave radiation (of any atmospheric "greenhouse" gas) is driven by the Planck function. Other so-called "greenhouse gases" including methane (some with larger absorption coefficients, but all with significantly less concentration) have their intensity quickly transferred upward and depleted by the same strong Planck function intensity change that applied to CO$_2$ and H$_2$O.

The trace gas methane (CH$_4$) has been receiving more attention than deserved lately and requires further discussion. It has been said that it is 20 times more efficient an absorber than CO$_2$; but this is only true in the two narrow bands of methane at 3.3 μm and 7.5 μm. Moreover, in solving for the possible impact on climate requires the solution of the Schwarzschild equation as shown earlier. This is where the concentration comes in and methane has no impact. The atmospheric concentration of CO$_2$ is ~ 410 ppmv and the concentration of methane is 1.7 ppmv. The ratio is approximately 240, so after the factor of 20 difference in absorbance, the truth is a factor of 12 in favor of CO$_2$. Further, the Planck radiation intensity decreases faster with height at the methane wavelengths compared to that of CO$_2$ at 14.9 μm (this is clearly seen in Fig 11.2). Finally it has been pointed out that the absorbance of water vapor would dominate over that of methane at those two narrow bands of methane mentioned above. *Methane has no impact on climate!*

## References: Chapter 11


1. Peixoto, J. P. and A. H. Oort, 1992: *Physics of climate*. American institute of Physics, New York, 522 pp.
2. Rothman, L. S., D., et al, 2009: The HITRAN 2008 molecular spectroscopic database. *Journal of Quantitative Spectroscopy & Radiative Transfer*, **110,** 139-204. Data provided by Pacific Northwest National Laboratory.
3. Data provided by Pacific Northwest National Laboratory (same as that provided to HITRAN).
4. Liou, K. N., 2002: *An introduction to atmospheric radiation*. Academic Press (an imprint of Elsevier Science), 583 pp.
5. Houghton, G. H., 1985: *Physical meteorology*. MIT Press, 442 pp.
6. Fleming, R. J., 2018: An updated review about carbon dioxide and climate change. *Environmental Earth Sciences*, **77,** 1-13.
7. The radiative Flux at level Z is $\Pi$ x B(Z) where the constant Pi comes from the hemispheric integration over all angles below the plane $Z = Z_i$. When comparing the ratio of Flux values at two different levels; e.g. the surface to that at 16 km, the value of Pi has no impact as $\Pi$ x B(SFC) / $\Pi$ x B(16 km): the $\Pi$ cancel. However the actual remaining flux at 16 km from Table 11.7, the $3^{rd}$ column is ($\Pi$) (0.3257) = (3.1416) (0.3257 = ~ 1.023 -- still effectively transparent. The smaller coefficients *are transparent,* so the remaining trivial heat at 16 km is transmitted to space.

CHAPTER **12**

# Why the Climate Does Change

## The Solar Influence on Climate Change

The temperature observations and the $CO_2$ correlations in Chapters 3, 5 and 7, with the detailed radiation calculations of Chapter 11, reveal that CO does not cause *climate-change*. The Sun is an obvious source of *climate-change* providing in one hour the amount of energy mankind uses in one year[1]. However, the insolation of the Sun is not the cause of *climate-change* as defined here (though insolation from two Milankovitch orbit factors influenced a hemispheric climate-change concerning ice volume as was discussed in Chapter 6). Chapter 7 introduced other solar causes and now is the point in our time travel to reveal the details of the *solar magnetic field influences*.

The solar system formed 4.6 billion years ago from the gravitational collapse of a spinning mass of hydrogen and recycled star dust. As the spinning gas and dust flattened into a disc, aggregates circling the star formed into planets including Earth. The vast majority of the solar system's mass resides in the Sun. Those components of the Sun that affect the Sun's magnetic field are described in this Chapter. Significant progress has been made in understanding the magnetic field of the Sun, other processes involving the Sun, and on the combined impact of these with cosmic rays[2].

119

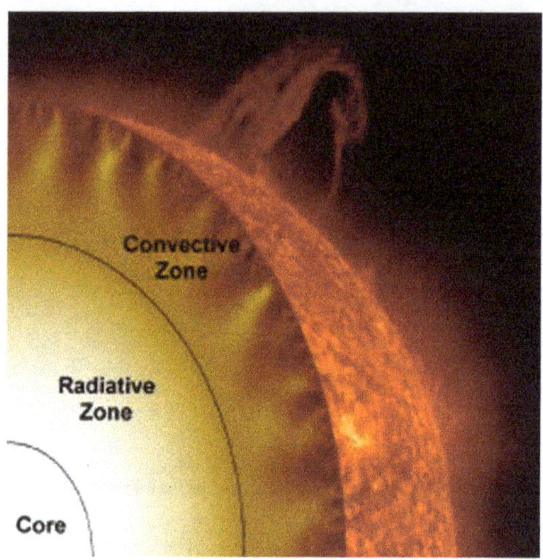


Figure 12.1  Solar interior

The Sun's interior has four different regions with unique processes in each (see Fig 12.1)[3]. The first is the **core**, the innermost 25%, where nuclear reactions consume hydrogen, produce helium and release energy to the surface as visible light. The temperature at the center of the Sun is about 15,000,000 °C. This temperature decreases toward the outer edge of the core.

The *radiative region* extends outward from the outer edge of the core to the interface layer or tachocline (a thin layer between the *radiative zone* and the *convective zone*). *This thin layer* plays an important role in generating the Sun's magnetic field. This will be discussed in greater detail a little later. The change in flow velocities across this thin layer (shear flows) can stretch magnetic lines of force and make them stronger. The change in the flow velocities give the layer its name – tachocline (tacho – borrowed from the Greek meaning "speed").

The *convective zone* covers the final 30% of distance (210,000 km) from the center of the Sun to the visible surface, the Photosphere. This convection follows the same principles as in atmospheric convection. Material moving upward which is warmer than its surroundings will continue upward. These solar convective motions carry heat *rapidly upward.* The fluid cools as it rises – the temperature at the upper visible surface has cooled to 5,700 K. The convective motions are visible at the surface as granules and supergranules and are shown in the photosphere in figures to follow.

The *Photosphere* is the visible surface of the Sun. it is a layer of gas about 100 km thick – quite thin compared to the ~ 700,000 km radius of the Sun. It does not rotate rigidly like a solid planet – it is a gas. The Sun's equatorial region rotates faster (24 days for a complete rotation) compared to the polar-regions which rotate once in ~ 30 days.

The photosphere reveals sunspots (see Fig 12.2) which appear as dark spots on the surface of the Sun. Temperatures in the dark centers of the sunspots are only 3,700 K compared to the 5,700 K for the surrounding photosphere. Sunspots are magnetic regions of the Sun with magnetic field strengths that are thousands of times stronger than the Earth's magnetic field.

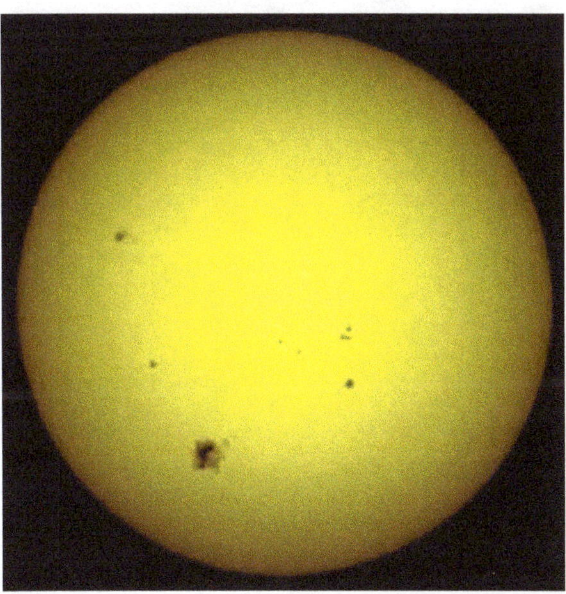

Figure 12.2  A few sunspots

Faculae (not shown here) are bright areas – usually most easily seen near the limb, or edge of the solar disk. These are also magnetic areas, but the magnetic strength is concentrated in much smaller bundles than in sunspots.

*Granules* are smaller cellular features (~ 1000 km diameter) that cover the entire surface of the Sun except for those areas covered by sunspots see Fig 12.3.

Granules are the tops of convective cells where hot fluid rises up from the interior in the bright regions, then spread out across the surface, cool, and then sink inward along the dark lanes. Individual granules last for only about 20 minutes, but the granulation pattern keeps rejuvenating itself with old granules being pushed aside by new ones.

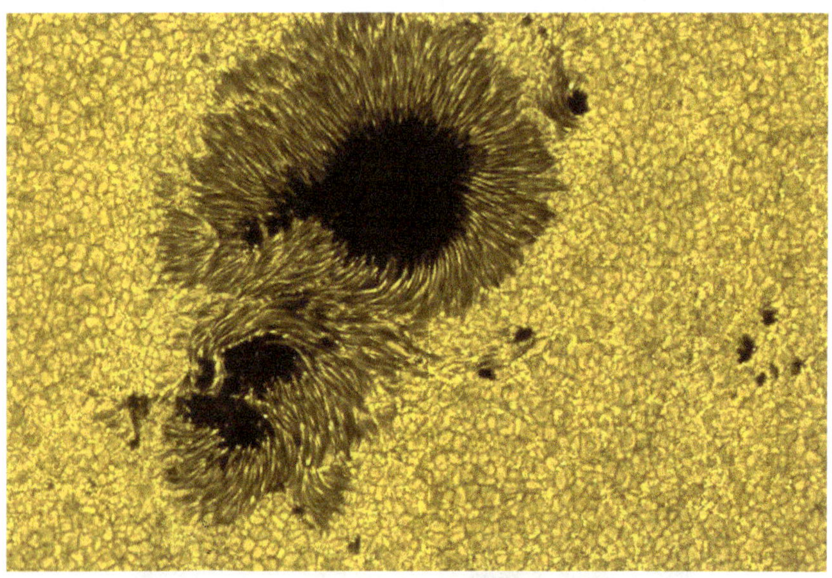

Figure 12.3  Close-up of sunspots and background of granules

*Supergranules* are much larger versions of granules (~ 35,000 km diameter) – see Fig 12.4. These convective elements also cover the entire Sun (except for areas covered by sunspots) and are continually evolving—individual elements last for a day or two. These are best seen with measurements of the Doppler shift (with light moving toward the viewer shifted in blue, and shifted away from the viewer in red).

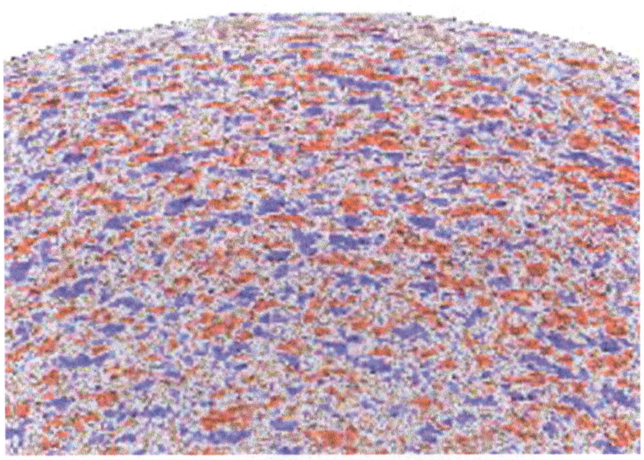

Figure 12.4  Field of the Sun's supergranules

Fluid flows observed in supergranules carry magnetic field bundles to the edges of the cells where they produce the chromospheric network of magnetic fields elements. The chromosphere is briefly discussed as part of the Sun's Corona.

The chromosphere is an irregular shaped layer above the photosphere where the temperature rises from 6000 °C to about 20,000 °C. At these temperatures hydrogen gives off a reddish light that reveals prominences that project above the rim of the Sun during total solar eclipses.

The *Corona* is the Sun's outer atmosphere. The coronal gases are heated to temperature greater than 1,000,000 °C – the cause of this high temperature is not yet certain. These high temperatures strip the electrons from hydrogen and helium, the two lightest and dominant elements of the Sun. These protons and electrons form a plasma that makes up the solar wind.

The corona shines brightly in x-rays because of the excessive temperatures. However, the cool solar photosphere (6000 °C) emits very few x-rays which allow the viewing of the corona across the disk of the Sun when viewed with x-rays. This showed scientists the existence of the "coronal holes" (see Fig 12.5). The "fast" solar wind (discussed below) is thought to originate from the coronal holes. The physics that gives the solar wind its speed is not yet clear – but will be soon, with the future NASA satellites coming to directly observe the Sun's details[4].

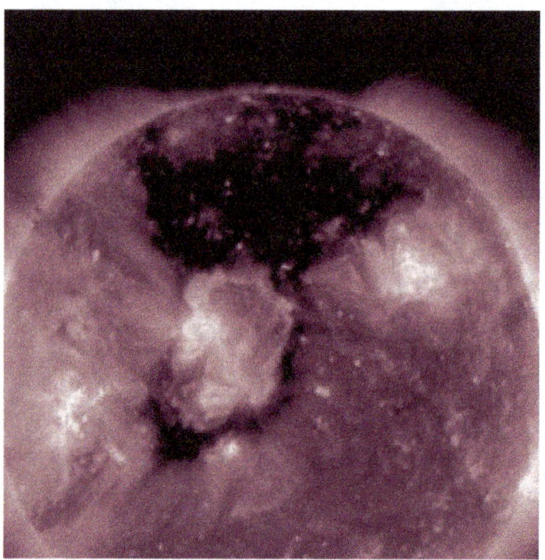

Figure 12.5  Coronal hole

Fig 12.5 indicates a coronal hole visualized in extreme ultraviolet light (invisible to the naked eye), but colorized in purple for easy viewing[5].

### The Solar Wind and the Sun's Magnetic Field

Energetic solar flares increase the Sun's ultraviolet radiation by 16%; and in Chapter 7 there was the reminder that the total magnetic flux leaving the Sun, dragged out by the solar wind, had risen by a factor of 2.3 since 1901[6]. The solar wind is a stream of charges particles, a plasma of matter in which electrons and protons have been separated, creating the hot mixture of such charged particles. The solar wind is composed primarily of hydrogen (95%), helium (4%), and carbon, nitrogen, oxygen, neon, magnesium, silicon and iron (~1%). These are all positive ions which implies that they have lost their electrons due to the high temperature of the Sun.

Magnetic fields are created when charged particles move. Currents flowing through the plasma of the solar wind give rise to a large scale magnetic field. The Sun's magnetic field causes sunspots, solar flares, and coronal mass ejections. Only a few details remain on how the Sun's magnetic field is *precisely* generated -- simulations are close, but a simple summary will follow.

The plan for the remainder of this Chapter is to discuss two quite different aspects of the Sun's activity (the variation of the solar sunspots in space and time) and the motion of the Sun about the barycenter (center of mass) of the solar system. Both of these activities relate to the Sun's magnetosphere and solar wind, therefore are related to the Sun's protection of the Earth from cosmic rays – affecting potential *climate-change* on Earth.

## Sunspot Variability

The Sun can be considered as a plasma of positive ions and negative electrons that can be treated as a continuum fluid with the equations of magnetohydrodynamics (MHD). One combines Maxwell's equations with the equations of fluid mechanics to drive the basic equations of MHD. In MHD, a plasma is a very good conductor of electricity, though large scale electric fields are not expected in the plasma, however, currents flowing through the plasma can give rise to *large scale magnetic fields.*

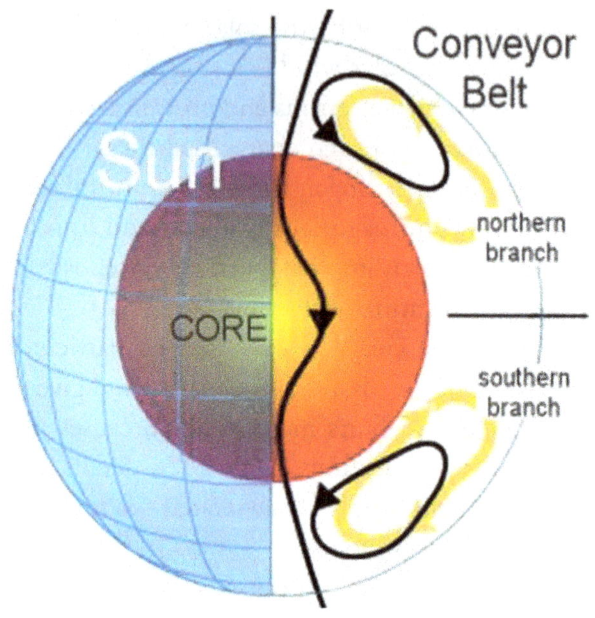


Figure 12.6  Sunspots form and move as if on a conveyor belt

The length of a sunspot cycle can vary from 9.8 to 12 years, (another source has 8 to 14 years) but the average period is 11.2 years. The magnetic dipole associated with the sunspots changes polarity at the sunspot maximum, thus the magnetic period is approximately 22.5 years.

The Sun's angular velocity about its spin axis is 20% higher at the equator than at the poles. This **differential rotation** would stretch out the magnetic field lines in the toroidal direction (i.e., the direction with respect to the Sun's rotational axis) – see Fig 12.7.

The left side of Fig 12.7 indicates an initial poloidal (meridional) magnetic field line. The right side indicates the stretched line in the toroidal direction. The magnetic field equation has a diffusion term (required in laboratory plasmas, but neglected in large scale astrophysical plasmas like the Sun; thus, the Sun's magnetic field is "frozen" in the plasma and moves with it. This adds to the strength of the toroidal field[7].

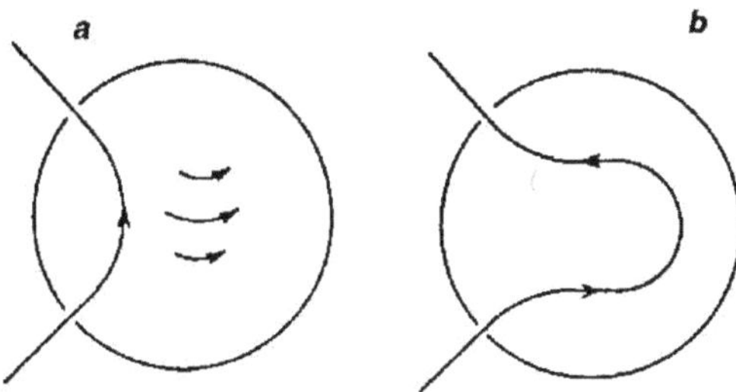


Figure 12.7  Production of a strong toroidal magnetic field

There was a brief discussion about the energy transported upward beneath the Sun's surface in the convection zone. However, to understand why the magnetic field is concentrated in structures like sunspots rather than being spread out more evenly requires a further explanation.

The nonlinear theory of convection in the presence of a vertical magnetic field reveals that the shared region gets separated into two kinds of areas. In certain areas the magnetic field is excluded and vigorous convection occurs. In other areas the magnetic field gets concentrated and the tension of the magnetic field lines suppresses the convection in these regions – this is where the sunspots are formed --- where the magnetic field is concentrated and the surrounding convection is suppressed. Then, since heat transport upward is greatly reduced by the suppressed convection, sunspots are cooler and look darker than the surrounding regions[7].

Further interaction within the convective region would keep vertical magnetic field lines in bundles throughout the convection zone. Such bundles of magnetic field lines are called *flux tubes* and are aligned in the toroidal direction in regions of strong differential rotation. If part of a flux tube rises through the surface via magnetic buoyancy as shown in Fig 12.8, then one has two sunspots of opposite polarity at nearly the same latitude – known as a bipolar sunspot.

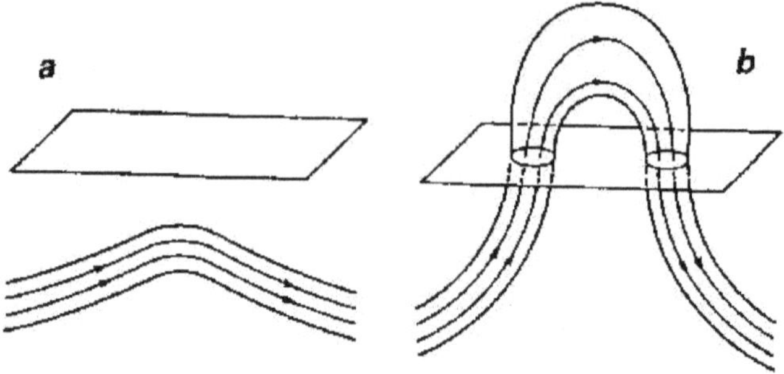


Figure 12.8  Magnetic buoyancy of a flux tube

Fig 12.8 indicates the nearly horizontal flux tube under the solar surface on the left of the diagram, and its upper part (on right) after magnetic buoyancy has caused it to rise above the solar surface. This is the beginning of the complete solar dynamo[7].

The process begins with the differential rotation (Sun's angular momentum -- stronger at the equator than at the poles) concentrated in the tachocline at the bottom of the convective zone. See the lower left of Fig 12.9 (a cartoon as a guide). Already seen in Fig 12.7, an initial poloidal (meridional) field line is stretched into forming the toroidal field by the differential rotation.

Then this toroidal field rises due to *magnetic buoyancy* (only when the magnetic field is as strong as $10^5$ Gauss can the magnetic buoyancy overcome the Coriolis force (from the Sun's rotation) and have flux tubes emerge at low latitudes) to produce bipolar sunspots at the solar surface (seen in Fig 12.8) and visualized by the dark upward arrows in Fig 12.9.

Then there is the action of the Coriolis force on the rising flux tubes. When the tilted bipolar sunspot pair decays: the polarity of the leading sunspot gets more diffused in the lower latitudes and the polarity of the following sunspot gets more diffused in the higher latitudes. This gives rise to the poloidal field at the solar surface – this sequence is called the Babcock-Leighton process.

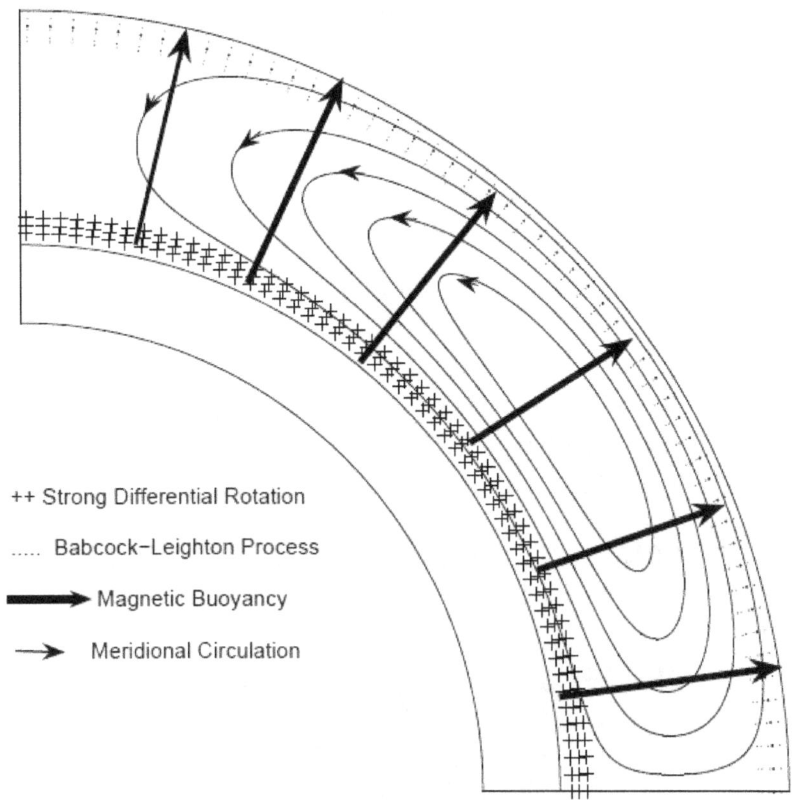


Figure 12.9  A cartoon explaining how the solar dynamo works

A tilted bipolar sunspot pair can be viewed as a conduit through which a part of the toroidal field ultimately gets transformed into the poloidal field[7]. This process occurs at the top of the convection zone – visualized by the dots at the top of Fig 12.9.

The poloidal field becomes part of the meridional circulation which has also been carrying the magnetic fields of dead sunspots from past granules/supergranules (knots of magnetism generated by the Sun's inner dynamo shown in Figs 12.3 and 12.4) poleward at a speed of ~ 20 m/s.

The meridional circulation exhibits a continuous flow from the equator to the poles: first to the polar region, and then underneath the surface down to the tachocline where the flow is equatorward at a pace of 1-2 m/s, where it is then stretched by the differential rotation – thus completing the cycle[7]. The return flow at the denser bottom of the convection zone takes approximately 11 years. A look at the complete cycle over several time periods is contained in Fig 12.10.

Scientists can monitor the sunspots as they move from high to low latitudes in the "Butterfly Diagram" of Fig 12.10. Those involved in simulating the solar dynamo and striving to understand the non-linear physics of the Sun have succeeded in reproducing a fair numerical representation of Fig 12.10[7]. A recent review of solar dynamo modeling is provided in the following reference[8].

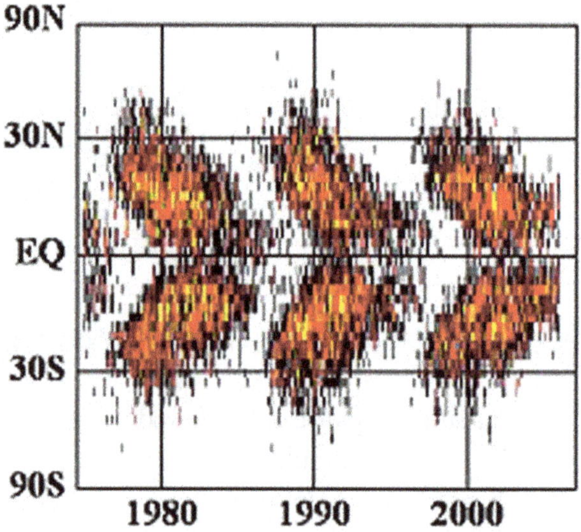


Figure 12.10  Sunspot movement from higher to lower latitudes

## Motion *of the Sun about the* solar system center of mass

The work of (Usoskin, et al. 2007) provides a history of Grand Minima since 9500 BC[9]. There have been 27 Grand Minima with various durations (from 30 to 110 years), with a tendency for Grand Minima to cluster with a quasi-period of about 2400 years. There were **not** significant *climate-changes events* with **all** of these Grand Minima.

A significant improvement in determining *which Grand Minima are important for climate-change* came with the work of (Sharp 2008) using information of the Sun's motion about the SSB (the solar system barycenter)[10]. The SSB is constantly changing position – primarily depending upon the location of the four massive planets in their respective orbits. The mass of the four planets: Jupiter (1.90 x $10^{27}$ kg), Saturn (5.68 x $10^{26}$ kg), Neptune (1.06 x $10^{26}$ kg) and Uranus (8.68 x $10^{25}$) – as well as the mass of the Sun itself, account for the SSB position.

Landscheidt[11] argued that the *orbital angular momentum of the Sun* with its motion about the SSB would add to the *Sun's spin angular momentum* (AM) (1.7 x $10^{48}$ g $cm^2$ $s^{-1}$)[10]. The Sun's orbital angular momentum (L) can range from near zero to 4.3 x $10^{47}$ or ~ 25% of the Sun's spin angular momentum – enough to make a *significant* impact on the Sun's inner dynamo (AM + L)

The orbital angular momentum (L) of those four massive planets are given by the formula: L = $2\prod$ m $r^2$ / P -- where m is the mass of the planet, r is the orbital radius of the planet from the Sun, and P is the orbital period in seconds. The results are in the following units [g $cm^2$ $s^{-1}$] with the numerical results: Jupiter (1.9 X $10^{50}$), Saturn (7.8 x$10^{49}$), Neptune (2.5 x $10^{49}$) and Uranus (1.7 x $10^{49}$). The Sun has orbital angular momentum as it moves about the SSB (though this distance is relatively small). This value ranges from near zero to 4.3 x $10^{47}$ g $cm^2$ $s^{-1}$.

Fig 12.11 shows the Sun in the geometric center of the solar system. The SSB can range from being near that geometric center of the solar stystem to being outside that geometric center by as much as 2.2 solar radii[10]. Note the position of the Sun in years 2029 to 2031 which focus on those years and year 2030 which will come up in the last Chapter.

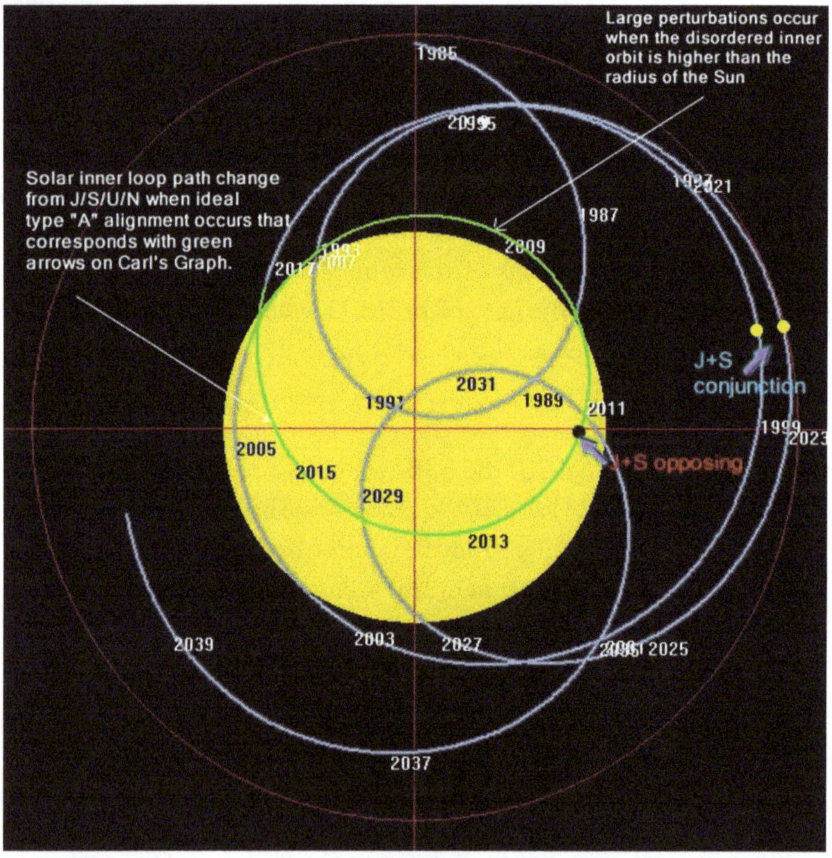


Figure 12.11  Motion of the Sun about the SSB: positions are indicated by years

Sharp furthered the work of Landscheidt and used the Jet Propulsion Laboratory DE405 ephemeris data and with the help of Carl

Smith[12] who also produced the results in Figs 12.12 and 12.13. His C-14 data in Fig12.12 is from Stuiver, et al.[13]. The results confirm the reason for the Medieval Warming (Grand Maximum) and the Little Ice Age (1300 to 1850) with its separate Grand Minima.

The negative delta C-14 data indicate the *lack of cosmic rays* (the Medieval Warming and the Modern Warming). The positive delta C-14 indicate the abundance of cosmic rays that represent the "quiet Sun" and the general cold periods of all the Minima, including those of the very cold Little Ice Age with its two separate phases ~ 1300 to 1550, the first, and the colder second phase 1550 to 1850. The 70 year Maunder Minimum was the coldest period – centered in 1680.

The synodic period $[T_S]$ (two successive conjunctions of the same bodies) of two planets 1 and 2 is given by $1/T_S = 1/T_1 - 1/T_2$ (with $T_1 < T_2$). The sidereal periods for Uranus and Neptune are 84.02 and 164.79 years respectively. This gives $T_{UN} = 172.42$ years. This is the main driver seen in the AM curves of solar grand minima and cosmic ray intensity over the past 9400 years.

When the SSB is far from the geometric center, the angular momentum is high – adding to rotational angular momentum of the Sun and increasing the strength of the solar dynamo. When the Sun is close to the center, this orbital AM of the Sun is near zero and the magnetic field weakens – as in Fig 12.11 and Fig 12.12. During each conjunction of Uranus and Neptune (every 172 years on average), Jupiter and Saturn have multiple *oppositions* (with a 20 year synodic period).

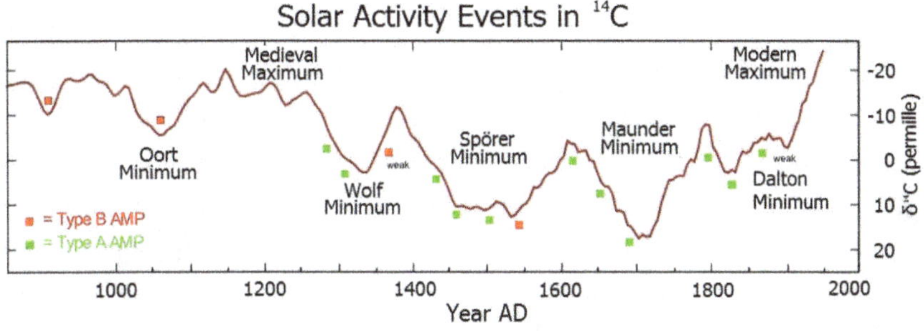


Fig 12.12  Solar activity from Sharp

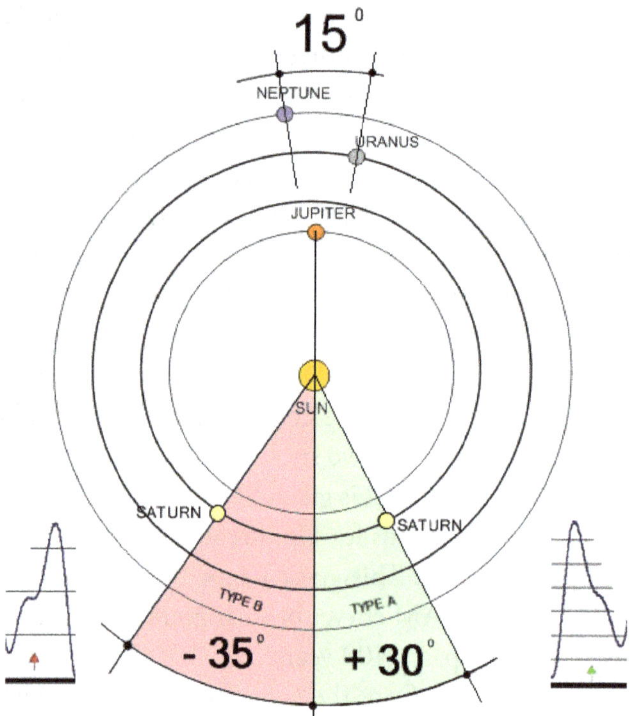



Figure 12.13  Typical planetary positions for all Type A and B events

A shallow inner loop is evident when Jupiter and Saturn are in *opposition*. Type A angular momentum perturbations (AMP) events have a major impact on the inner loop trajectory of the Sun as it orbits the SSB. A significant solar minimum involves the giant planets in their relationship with the Sun and as depicted in Fig 12.13 -- *Uranus, Neptune and Jupiter together* and Saturn opposite the Sun.

McCracken, et al.[14] extended Sharp's results back through the last 9400 years. The data sources for cosmic rays were the cosmogenic radionuclides Be-10 and C-14 from ice core records and tree rings, respectively. The data over the entire record confirms Sharp's results, and their statistical analysis found periodicities near 350, 510 and 710 years which closely approximate integer multiples of half the $T_{UN}$ synodic period: $T = (T_{UN} / 2) N$ years with N = 4, 6, and 8. Using combinations of these periods one could *approximate* the transition between the various warm and cold periods observed in the past few 1000 years.

The McCracken data over the entire record confirms Sharp's results and provide a number of further independent indications that there is a strong empirical correlation between the motion of the Jovian planets, the cosmic-ray intensity, and solar activity.


## Earth's Major Ice Ages

It is now appropriate to discuss in far more detail, the connection of the cosmic rays reaching the Earth's atmosphere *to produce enough cloud cover to change the Earth's albedo.* Providing the degree of cooling for the Little Ice Age is one thing; — providing sufficient cooling decade after decade for the Ice Ages to form is quite another. This information only came together in the last dozen years so its introduction now fits the schedule of our travelogue. *Appendix E* will help.

The Milky Way Galaxy is nearly as old as the universe itself at 13.8 billion years. It is immense – 100,000 – 200,000 light-years in diameter. There are four spiral arms of stars surrounding the Galactic Center – where there exists a super massive black hole (SMBH). Our solar system lies 27,000 light-years from the Galactic Center, on a small local spur (Orion Arm) between the Sagittarius and Perseus spiral Arms (follow line 180 up in Fig.12.14 from NASA).

The Galaxy exists in the form of a disc with a central bulge with a diameter of 12,000 light-years. The force of gravity acting between the stars generates waves of *dense and less-dense matter.* These density waves perturb the interstellar gas producing relatively dense clouds from which new stars are born. Massive bright bluish-white stars, with relatively short lives populate the arms.

Small stars like the Sun live long enough to orbit around the center of the Galaxy many times. The journey about the Milky Way Galactic Core takes approximately 230 million years[2].

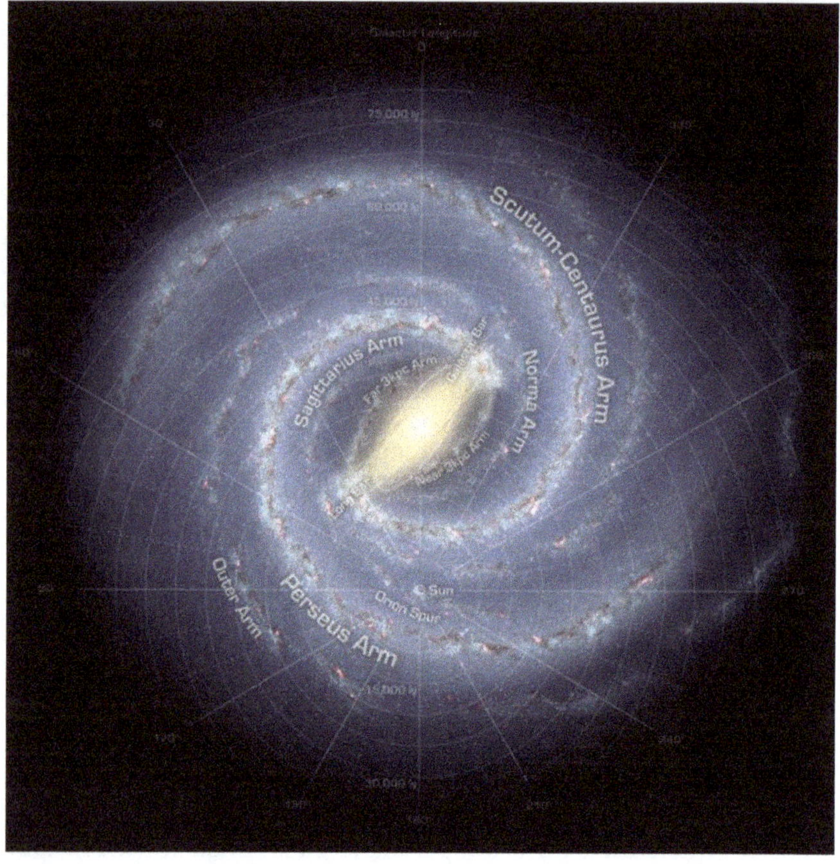


Figure 12.14  Milky Way spiral arms

The solar system moves faster through the Milky Way galaxy than the rotation of the spiral arms, thus it is repeatedly running through the arms. The spiral arms contain many cosmic rays!

The solar effects on the Earth's climate are now undeniable and the essential assets of the Sun contributing to these climate changes have been identified. Now the Sun's motion as it travels with the solar system about the Milky Way galaxy will be discussed. Clearly a significant force must be identified to cause the Major Ice Ages the Earth has experienced in the past.

The solar wind drags the magnetic field of the Sun with it. The solar wind has speeds of ~ 350 or 750 km/s (depending on the source region of the Sun) and this makes the heliosphere reach as far as its limit (when interstellar gas can finally resist it) at five times the distance of Neptune, the most remote of the four giant planets. How will the strength of the Sun's magnetic field deal with the trip around the Milky Way galaxy?

The explosion of massive stars at the end of their lives produce cosmic rays. After a million years a supernova has dissipated its energy and becomes a neutron star. At any one time there are thousands of supernova remnants dispersing cosmic rays across the Milky Way galaxy. Twenty years ago Svensmark[15] was the first to suggest that cosmic rays could change the climate.

Cosmic rays reaching the Earth's upper atmosphere encounter a major change – the high speed protons and nuclei of heavier atoms come to a halt and produce secondary cosmic rays. These swarms of secondary rays are fast moving particles released in atomic and nuclear interactions. These secondary particles go on to have further encounters of their own, so that the sum of the interactions leads to further showers of millions/billions of particles.

The electrons in the cosmic rays and in the atmosphere are too light to have any impact. However, there are *charged particles created within these showers* that do have an impact; this is the *muon* similar to an electron in every respect except that it is 200 times heavier. Svensmark and his team provided the proof of the cosmic rays accelerating the formation of cloud growth in 2006 – nine years after his first scientific paper. See *Appendix E* for the details.

Neighbors to the Milky Way include the Large and Small Magellan Clouds, and the Andromeda galaxy; together with some 50 other smaller galaxies this cluster is known as the Local Group. Farther out is the Virgo Supercluster which includes the Local Group and another 100 galaxy groups – this has a 100 million light-year diameter. This immense region is a source of potential cosmic rays interacting with Earth.

Two other key scientists enter this picture having studied the interaction of cosmic rays with the spiral arms of the Milky Way Galaxy – even before Svensmark's theory was proven. Jan Veizer, a geologist from Canada, reconstructed a temperature record for the last 500 million years using isotopes in fossilized seashells – he found a warming/cooling cycle every 135 million years[16].

Later Nir Shaviv, an astrophysicist from Israel, visited with Veizer in Canada and relayed his discovery of cosmic rays being received on Earth on a similar (137) million year cycle as the solar system passes through the bright arms of the Milky Way Galaxy. The two subsequently teamed up and published a paper concluding that 75% of the Earth's temperature variability over the past 500 million years was due to the cosmic rays striking Earth as the solar system passed through the bright spiral arms of the Milky Way Galaxy[16].

Shaviv found a way to check on cosmic ray variability in the distant past. He analyzed radioactive data in meteorites -- fragments from colliding asteroids (some with lumps of iron) which have been orbiting around the Sun for hundreds of millions of years. Cosmic rays produce radioactive potassium, on these particular orbiting asteroids. Potassium (atomic weight 39) gains a neutron from cosmic rays to become Potassium-40. K-40 has a half-life of 1.25 billion years. Eventually some of these asteroid fragments fall to Earth as iron meteorites.

One might gauge how long these iron meteorites had been orbiting around the Sun from the amount of radioactive K-40 they contained in proportion to other stable atoms, but *variations* in the intensity of cosmic rays experienced in the solar system would affect the accuracy of the results. The apparent ages of iron meteorites bunch unnaturally when cosmic rays were scarce. Obtaining independent dates of cosmic ray impact required Shaviv to exclude cases where the character and ages of the meteorites were too similar. Finally, using just 50 time-independent iron meteorites ranging from up to a billion years in age, he was able to determine details of cosmic ray intensity.

His revised estimate of the cosmic ray cycle was 143 million years plus or minus 10 million years. Meanwhile geologists around the world had been working over the past half century, recognizing alternating periods of hot-house periods and Ice Age periods, and trying to refine the dates of these events. Shaviv's best fit to those determined climate periods was 145 million years – close to his cosmic ray cycle[17]. It will be shown that every visit to the spiral arms produced an Ice Age without – exception. Shaviv expressed his thoughts in a quote in Svensmark and Calder's book[2]:

"The variations in the cosmic ray flux rising from our galactic journey are ten times larger than the variations due to solar activity, at the high cosmic-ray energies responsible for ionizing the lower atmosphere. If the Sun is responsible for variations in the global temperature of about 1 degree Celsius, the effect of the spiral-arm passages should be about 10 degrees. That is more than enough to change the Earth from a hothouse where temperate climates extend to polar-regions, to an ice-house with ice-caps at the poles, as is the case today. In fact, the spiral-arm effect is expected to be the most dominant driver of climate changes over periods of hundreds of millions of years".

There is another motion of our solar system as it traverses the Galaxy. Here we quote from the Svensmark / Calder book: "While the Sun orbits the Milky Way, it also rises and plunges and rises again repeatedly, through the disc of stars that surrounds the central bulge of the Galaxy. Cosmic rays concentrate in the disc because the magnetic field that guides them is held in place by the gravity that keeps stars and gas clouds confined to the disk.

The cosmic rays are more intense for the Earth whenever the Sun crosses the mid-plane, whether going up or down, which happens at intervals of about 34 million years". The image to think of is a 'Merry-Go-Round' with the wooden horses going up and down as it rotates. These variations occur whether the solar system is inside or outside a *spiral arm*.

The charged particles from an exploding supernova propagate to the outer edge of the Galaxy, the galactic corona; then follow magnetic field lines back toward the center of the Galaxy where they provide fuel for the next generation of stars. This represents a continuous process of star creation and destruction within the *spiral arms*. However, our Galaxy contains ~ a billion solar masses of gas which are *available to form stars*, yet it produces just *one solar mass of new stars per year*[18]. Accounting for this inefficiency is a major challenge for astrophysics. The internal motion of the galactic gas is chaotic and highly turbulent.

This turbulence, gravity, the magnetic fields and nonturbulent motion all interact to create this phenomena, but the relative roles have not been fully determined[18]. Ross opines that this allows mankind to visualize our universe more clearly[19].

The following paragraphs will provide the reader with a tour about the Milky Way Galaxy where the Ice Ages will be are encountered as the solar system encounters the spiral arms. A few facts may be thrown in about biological progression of life forms during the journey. This tour will cover the *last* 500 million years of Shaviv's billion year record. The format for each solar system visit will provide: the timing of the ice age relative to the era, the name of the Ice Age, and the spiral arm that that solar system penetrated. Then we will have something to say about the *first* 500 million years of his record.

The strongest, coldest Ice Age occurred at the end of the Proterozoic ion in the Neoproterozoic era. The Cryogenian Ice Age progressed from 750 to 580 million years ago (MYA). The solar system had just confronted the Sagittarius Arm of the Milky Way. There were long periods of alternating glaciations and interglacials (when the sea surface temperature was 40°C -- sea level rose and fell 600 meters. Life on Earth then was only bacterial[2].

Following this Ice Age, the warm Cambrian Period from 540-520 MYA produced an explosion of life forms. Cosmic rays were few and sea level was high. New life formed on the continental shelves. A phyla which included fishes emerged along with all other animals with backbones[2].

The next Ice Age occurred in the Ordovician Period, the Andrean-Saharan from 460 to 430 MYA. The solar system intercepted the Perseus Arm. This was a relatively short ice age. It was followed by the warm Silurian Period where the first plants and animals lived on the land[2].

The third major Ice Age began in the Carboniferous Period and extended into the Permian Period as the Karoo Ice Age from 350 to 280 MYA. The solar system confronted the Norma Arm. The values of $CO_2$ concentration had plunged to values near to those seen today. The cold oceans had adsorbed the large amounts of $CO_2$. [See Fig. 3.1 for the low values of $CO_2$ and temperature.]

Following this ice age, the warmth returned to the entire Triassic and early Jurassic Periods as the Earth moved between the spiral arms – the cosmic rays were significantly reduced and the Earth's oceans and atmosphere warmed. The oceans outgassed large amounts of $CO_2$ and the concentration increased again to 4-5 times today's value of approximately 400 ppmv. The era of maximum growth of the partial pressure of oxygen occurred from 250-200 MYA[20]. This accelerated life forms based upon oxidation of organic matter and further increased $CO_2$.

Table 12.1  Solar system visits through the Spiral Arms of the Milky Way Galaxy

| Name of the Ice Age | Period | Time extent (Mya) | Spiral Arm |
|---|---|---|---|
| Cryogenian | Neoproterozic | 750 to 580 | Sagittarius |
| Andrean-Saharan | Ordovican | 460 to 430 | Perseus |
| Karoo or Permo-Carboniferous | Carboniferous into Permian | 350 to 280 | Norma |
| Jurassic-Cretaceous Glaciation | Mesozoic | 150 to 132 | Scutum-Crux |
| Present | Tertiary-Quaternary | 50 to present Antarctica Ice -37 Mya N. H. – 2.7 Mya | Sagittarius to Orion (Local spur) |

The fourth Ice Age occurred in the Jurassic-Cretaceous Periods. It is called the Scutum – Crux Ice Age occurring from 150-132 MYA. This was a less severe high latitude Ice Age[1]. The solar system intercepted the Scutum-Crux Arm of the Milky Way. The latest Ice Age is the *Current* Ice Age. The solar system is back in the Orion Arm and there has been cooling for the past 50 million years[2] with alternating cold and warm periods (glacial and interglacial). Antarctica has had ice sheets for the past 37 million years.

The oldest *suspected, but not fully documented,* ice age is the Huronian which occurred 2.4 to 2.1 billion years ago. This straddled the large $CO_2$ rise and initial depletion referred to earlier. Geologists have generally agreed that an extremely strong Ice Age occurred in the Paleo-Proterozoic era between 2,400 and 2,200 million years ago. Glaciation was at sea level and in the tropics where the continents had merged into a single land mass[1].

A scientific surprise resulted from the Infrared Astronomy Satellite in 1983 indicated galaxies far warmer than expected with extremely strong infrared signals. Later Europe's Infrared Space Observatory had examined hundreds of these ultra-luminous objects in detail and confirmed that the luminosity of these galaxies was due to extremely vigorous star formation. These galaxies are now called star-burst galaxies. The infrared rays result from warm dust produced by numerous explosions of massive short-lived stars – usually due to collisions between galaxies which can create shock waves that compress gas between the stars and produce new stars.[2].

Astronomers can check on star formation rates, but need the distance to a star to determine its age. Distances to stars became more accurate in 1997 with Hipparcos, Europe's star-mapping satellite. Several periods of new star formation were found in the Milky Way Galaxy – one of these occurred in the 2,400 to 2,200 MYA period. This was a direct link to the Huronian Ice Age of that period *and a solid confirmation of the solar/cosmic ray connection.*

A second very important historical fact is that in the long period from 2,000 MYA to 750 MYA there was over a billion years of very little star formation – and no glaciation on Earth during that period. Only at the extremes of those time periods did the two largest Ice Age events occur on Earth -- the Huronian Ice Age and the Cryogenian Ice Age (both now discussed previously). This *billion-year null event* lends further support to the theory[2].

A third important set of facts supporting the solar/cosmic ray connection involves a previous mystery of why the early *Sun radiated only 70% of its present sunlight,* and yet there are clear indications of liquid water on the Earth in very early times. Rocks from 3,800 million years ago in the Archean Eon show signs of water accumulation on ancient sea-beds.

The most logical reason of several that have been proposed, is the state of the albedo of Earth at that time. There were no ice sheets and there were very few clouds due to cosmic rays being shielded by that Sun – the albedo would have been very low – allowing the full intensity of the Sun's radiation to be received at the surface of Earth.

The reason for the cosmic rays being shielded is from the fact that the Sun had a rotation rate 10 times faster than today. Astronomers know this from the studying young sun-like stars as well as by theories of the Sun's internal history[2]. That would have created a *much stronger magnetic field,* a more intense solar wind keeping cosmic rays at bay and *allowing the full force of insolation.*

## References: Chapter 12

1. Plimer, I., 2009: *Heaven and Earth -- global warming: the missing science.* Quartet Books, Limited, London, 503 pp.
2. Svensmark, H. and N. Calder, 2007: *The chilling stars – a cosmic view of climate change.* Icon Books Ltd, Thriplow, Cambridge, UK, 268 pp.
3. https://solarscience.msfc.nasa.gov/interior. Web author Dr. David H. Hathaway. There are a variety of solar images made available by NASA on the internet that are included here in this Chapter.

4. Sokol, J., 2018: A place in the sun. *Science*, **361**, 441-445.

5. The special image of Fig. 10.5 from imagery of Sun captured by NASA's Solar Dynamics Observatory on May 17-19, 2016.

6. Lockwood, R., R. Stamper and M. N. Wild, (1999): A doubling of the Sun's coronal magnetic field during the past 100 years. *Nature*, **399**, 437-439.

7. Choudhuri, A. R., 1998: *The physics of fluids and plasmas: an introduction for astrophysicists.* Cambridge University Press. See also internet paper march 7, 2007. An elementary introduction to solar dynamo theory, www.physics.iisc.ernet.in/~arnab/lectures/

8. Strugarek, A.et al., 2017: Reconciling solar and stellar magnetic cycles with nonlinear dynamo simulations. *Science*, **357**, 185-187.

9. Usoskin, I., S. Solanki and G. Kovaltsov, 2007: Grand minima and maxima of solar activity: new observational constraints. *Astronomy & Astrophysics*, **471**, 301-309.

10. Sharpe, G. J., 2008: Are Uranus and Neptune responsible for Solar Grand Minima and solar cycle modulation? First in http// www.landscheidt.info.arvix.org/ftp/arvix/pa-pers/1005/1005.5303.pdf.

11. Landscheidt, T., 2003: New little ice age instead of global warming. *Energy and Environment*, **14**, 327-350.

12. Smith, C., 2007: Angular momentum graph. http://landscheidt. auditblogs.com/2007/06/01.

13. Stuiver, M., P. J. Reimer, E. Bard, J. W. Beck, G. S. Burr, et al., 1998: INTCAL98 ra3iocarbon age calibration, 24,000#0 cal BP. Radiocarbon 40, 1041#83.

14. McCracken, K.G., J. Beer and F. Steinhilber, 2014: Evidence for planetary forcing of the cosmic ray intensity and solar Activity throughout the past 9400 years. *Solar Physics*. **289**. DOI 10.1007/ s11207-014-0510-1.

15. Svensmark, H. and E. Friis-Christensen, 1997: Variation of cosmic ray flux and global cloud coverage – a missing link in solar-climate relationships. *Journal of Atmospheric and Solar-Terrestrial Physics, **59**, 1225-1232.*

16. Shaviv, N. and J. Veizer, 2003: Celestial driver of Phanerozoic climate? *Geological Society of America,* **13**. 4-10.

17. Shaviv, N. 2005: Cosmic rays and climate. *PhysicaPlus,* online magazine of the Israel Physical Society.

18. Federrath, C., 2018: The turbulent formation of stars. *Physics Today,* June, 39 -42.

19. Ross, H., 2008: *Why the universe is the way it is.* Baker Books, 40 pp.

20. Sorokhtin, O. G., G. V. Chilingar and L. F. Khilyuk, 2007: *Global warming and global cooling: evolution of climate on Earth.* Elsevier, Amsterdam, 313 pp.

to hardwood and softwood tree growth. The volume of growing stock of hardwood and softwood trees in US forests has grown by 49% between 1953 and 2006[5]. Fig 13.1 is an earlier assessment with similar numbers.

The Idso family: Sherwood[6] and his two sons Craig and Keith have been active in promoting the plant productivity responses of increased $CO_2$ since 1989. The importance of $CO_2$ on world food supplies[7] is extremely important in a growing world population and a colder climate coming.

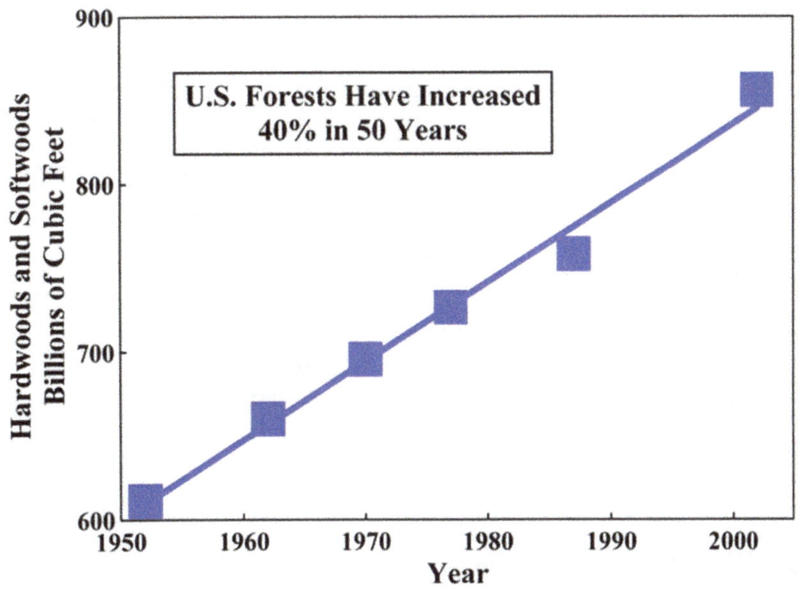


Figure 13.1  USA forest growth over a recent 50 - year period

Over a thousand peer-reviewed articles exist on the subject of biological enhancement of carbon dioxide enrichment[8]. Virtually all of these speak of the enhanced growth with additional $CO_2$ available. Doubling the level of carbon dioxide raises the net productivity of herbaceous plants by 30 to 50 percent and of trees and woody plants by 50 to 80 percent[9]. *Two further beneficial categories, important for changes that will come with further cooling, are discussed briefly.*

A study of soybeans grown at 700 ppm $CO_2$ displayed 10 to 25 percent reductions in total water loss while simultaneously increasing dry weight by as much as 33%. Elevated $CO_2$ significantly increased the *water-use efficiency* of the studied plants[10] – making them more drought tolerant.

A number of studies have verified the fact the $CO_2$ enrichment allows plants to withstand the rigors of low temperatures. A study of *the effects of chilling stress* under conditions of 360 versus 700 ppm $CO_2$ over a controlled three month period showed the higher concentration of $CO_2$ led to less stress[11]. This phenomena could provide a needed boost to sustain food production for a growing population in the coming years – especially when the climate turns cooler, which it will.

A general picture of past *climate-change* is that cold/dry conditions go together or otherwise warm/moist conditions occur. This can be expected from the fact that the warmer atmosphere can hold more water vapor, however, in past climate-change periods this has not always led to more precipitation available for mankind's use. Water vapor increase is limited by its strong feedback with precipitation. The atmosphere is like a wet sponge – the more it is squeezed, the more water comes out and then there is little left – water vapor increase has its upper limits.

Now we can stop worrying about $CO_2$ and global warming! What about air pollution which is often incorrectly associated with $CO_2$? Part of this error is made by those confusing $CO_2$ and CO (carbon monoxide) which is a toxic gas. $CO_2$ is not toxic.

Automobiles with internal combustion engines fueled by gasoline were a problem prior to 1975. But since then catalytic converters are in autos across the world. The latest are Three-way catalytic converters (TWC) that covert toxic gases and pollutants into valuable $CO_2$ and water ($H_2O$). Now virtually every country in the world has stringent vehicle emission requirements that in effect require three-way converters on gasoline-powered vehicles.

*Catalytic* converters are usually associated with exhaust systems in automobiles, they are also found on electrical generators, trucks, buses, motorcycles and locomotives. These applications were driven by health and safety regulations and government environmental regulations.

The TWC have three simultaneous tasks: (1) reduction of nitrogen oxides to nitrogen ($N_2$), (2) oxidation of carbon monoxide to carbon dioxide, and (3) oxidation of unburnt hydrocarbons (HC) to carbon dioxide and water. These three tasks are driven by the following chemical relationships[12]:

*Reduction of nitrogen oxides to nitrogen:*

$$2\,CO + 2\,NO \rightarrow 2\,CO_2 + N_2$$
$$2\,H_2 + 2\,NO \rightarrow 2\,H_2O + N_2$$

*Oxidation of carbon monoxide to carbon dioxide*

$$2\,CO + O_2 \rightarrow 2\,CO_2$$

*Oxidation of unburnt hydrocarbons (HC) to carbon dioxide and water*

$$Hydrocarbon + O_2 \rightarrow H_2O + CO_2$$

The *catalyst* is most often a mixture of precious metals. Platinum is the most active catalyst, and widely used, but has drawbacks of cost and unwanted additional reactions. Palladium and rhodium are two other precious metals used. Palladium is used as an oxidation catalyst and rhodium is used as a reduction catalyst.

Engines fitted with three-way catalytic converters are equipped with a computerized closed-loop feedback fuel injection system using one or more oxygen sensors for optimal performance. When a catalytic converter fails (the average life is about the same as the average car life ~ 100,000 miles driven) it is recycled into scrap and the precious metals inside the converter are extracted[12].

Carbon dioxide has no impact on air pollution from automobiles. Power plants also now have similar systems to extract harmful toxins and pollutants. There are a variety of solutions to reduce air pollution from coal and natural gas power producers of electricity.

There is another $CO_2$ issue that has been brought up by environmentalists. It is another scare tactic provided to the press for further fodder – the subject is ocean acidification -- which would *presumably lead* to the destruction of all the coral reefs.

This will be easily and soundly put down by the ideas expressed by the Australian Ian Plimer in his book[1].

The oceans have a greater storage capacity for $CO_2$ than the atmosphere and even more than that of the plant world. The oceans are constantly removing $CO_2$ to form carbonate sediments which eventually become carbonate rocks – containing 40,000 times more $CO_2$ than the atmosphere. The wind also pumps $CO_2$ into the oceans[13].

The oceans currently have an acidity measured as pH value 7.9 to 8.2. This measurement is higher than the neutral value (pH = 7) which implies that the oceans are alkaline (opposite of acidic). The pH scale ranges from 0 to 14. The pH value 6 is 10 times more acid than pH 7 and pH 5 is 100 times more acid than pH 7.

In order to acidify seawater from pH 8 to pH 6 would require an enormous amount of acid. Once there is acid present in the oceans, sediments, rocks and shells become very reactive. The reactions destroy acid and the oceans returns to their normal alkaline state.

If more $CO_2$ were added to the oceans then calcium carbonate would precipitate. Calcium carbonate $CaCO_3$ accounts for more than 4 % of the Earth's crust and is found throughout the world. The natural forms of $CaCO_3$ are chalk, limestone and marble, produced by the sedimentation of the shells of small fossilized coral over millions of years.

When $CO_2$ is dissolved in seawater, it is neutralized to bicarbonate by reacting with dissolved carbonate and borate in water and with calcium carbonate sediment covering much of the ocean floor. The geological record shows that shells do not dissolve, otherwise there would be no shelly fossils. The oceans are saturated with calcium carbonate to a depth of 4.8 km. If *any more* $CO_2$ were added to the oceans, then *more* calcium carbonate would precipitate.

The balance of $CO_2$ between the oceans and atmosphere we have today has not changed for thousands of millions of years[14]. This balance has not changed during times of intense sudden release of $CO_2$ from volcanoes. Increased volcanic production of $CO_2$ correlates very well with increased sedimentation of calcium carbonate from the oceans[15].

Rainwater is slightly acidic (pH 5.6), but by the time it runs over the ground and chemically reacts with minerals in soils and rocks, it enters the oceans as alkaline water. Soils contain more $CO_2$ than the atmosphere, and during weathering release a large amount of $CO_2$ that ends up in thousands of rivers and streams. The process of weathering has removed $CO_2$ from the atmosphere and soils for billions of years and storing it in rocks. [Perhaps as a bit of humor, Plimer points out that this *removal of $CO_2$ does not trigger glaciation.*] Plimer's final quote: "While the oceans have an excess of calcium, they cannot become acid."

## References: Chapter 13


1. Plimer, I., 009: *Heaven and Earth – global warming: the missing science.* Quartet Books, Limited, London, 503 pp.
2. Svensmark, H. and N. Calder, 2007: *The chilling stars – a cosmic view of climate change.* Icon Books Ltd, Thriplow, Cambridge, UK, 268 pp.
3. Sorokhtin, O. G., G. V. Chilingar and L. F. Khilyuk, 2007: *Global warming and global cooling: evolution of climate on Earth.* Elsevier, Amsterdam, 313 pp.
4. http://www.webmd.com/lungs/how-we-breathe.

5. Alveraz, M., 2007: The state of America's forests. Society of American Foresters. Bethesda, MD, 76 pp.

6. Idso, S.B., 1989: *Carbon dioxide friend or foe?* IBR Press, Tempe, AZ.

7. Idso, C. D. and K. E. Idso, 2000: Forecasting world food supplies: The impact of rising atmospheric $CO_2$ concentration. Technology, **7** (suppl.): 33-56.

8. Singer, S. F. and C. D. Idso, 2009: *Climate Change Reconsidered.* The report of the Nongovernmental International panel on Climate Change. The Heartland Institute. 856 pp.

9. Serraj, R. et al., 1999: Soybean leaf growth and gas exchange response to drought under carbon dioxide enhancement. *Global Change Biology*, **5**, 283-291.

10. Kimball, B. A., 1983: Carbon dioxide and agricultural yield: an assemblage and analysis of 430 prior observations. *Agronomy Journal*, **75**, 779-788.

11. Schwanz, P., and A. Pelle, 2001: Growth under elevated $CO_2$ ameliorates defenses against photo-oxidation stress in poplar (Populous alba_x tremolo). *Environmental and Experimental Botany*, **45**, 45-53.

12. https://en.wikipedia.org/wiki/Catalytic_converter

13. Smith, S. D. and E.P. Jones, 1985: Evidence of wind-pumping of air-sea gas exchange based on direct measurements of $CO_2$ fluxes. *Journal of Geophysical Research*, **90**, 860-875.

14. Holland, H, D., 1984: *The chemical evolution of the atmosphere and oceans.* Princeton University Press.

15. Budyko, M, I. et al., 1987: *History of the Earth's atmosphere.* Springer-Verlag

CHAPTER **14**

# Future Research on Climate and Energy Issues

The United Nations has defined the length and domain of *climate-change as the multi-year change of the Earth's averaged global surface temperature.* It has been shown that this particular limited view of climate-change is not caused by $CO_2$ but by the *Sun's magnetic field, the motion of the Sun about the solar system barycenter, and by cosmic rays from space.*

Many climate scientists will have to change their direction. We encourage such scientists to tailor their research toward the *many other space/time scenarios of climate-change that need to be better understood to the point of providing trusted and meaningful information to the world's population.*

Only a few of these space/time concerns will be mentioned here. They may involve only Earth sciences, or may also involve extra-terrestrial forces interacting with Earth sciences. This will be the content of the first part of Chapter 14.

Another important issue exists -- crucial to the quality of life that we all seek. This is the proper integration and use of *various sources of energy* required to fulfill our needs. This is energy to keep us warm, fuel our transportation, and energy to produce the products we enjoy and have become to depend on. Now one can use energy from fossil fuels and other energy sources – in the least expensive manner for every uniquely different application. A discussion of these opportunities will be the content of the second part of Chapter 14

## Chapter 14 Future Research on Other Climate Change Issues

There was an international science program under the United Nations called the Tropical Ocean and Global Atmosphere (TOGA) program which was a ten-year (1984 to 1994) aimed specifically at the prediction of climate phenomena on time scales of months to years. It grew out of the important economic impact of events that had been referred to as El Niño and La Niña oscillations driven by atmosphere and Tropical Ocean coupled interactions.

The location of these oscillations are in the tropical Pacific Ocean. Atmospheric pressures force the usual steady equatorial trade winds driving warmed surface waters westward across the Pacific Ocean toward Indonesia. Sea level in the western Pacific is higher by 1-2 feet. When the pressure gradient subsides so also the trade winds. Waves of warm water move back from west to east.

The usual cold waters off the South American coast become warm and the normally excellent fishery resources are severely depleted. This aperiodic event does not occur every year, but when it does return, it often occurs around the Christmas season – hence the return of the warm waters was given the name El Niño -- Spanish for the "Christ child." Sea surface temperatures (SST), are measured in the Niño 3.4 region (5°N -5°S, 120° - 170°W), and are used to define the El Niño event and distinguish it from its counterpart the La Niña event.

When the waters off Peru are *colder than normal* the condition is referred to as *La Niña.* Things change quite differently on both sides of the Pacific Ocean when either of these circulation cycles occur – both in climatic atmosphere and oceanic conditions, and especially in terms of economic factors. The Table 14.1 indicates *El Niño, La Niña,* and *neutral events* [1] over the period from 1975 to 2018 – where the events were over periods of consecutive years, e.g. 75'- 76'.

Table 14.1  Years of El Niño, La Niña and Neutral events

| | | El Niño | | Neutral | La Niña | | |
| Weak | Moderate | Strong | V. strong | | Weak | Moderate | Strong |
|---|---|---|---|---|---|---|---|
| | | | | | | | 75-76 |
| 76-77 | | | | | | | |
| 77-78 | | | | | | | |
| | | | | 78-79 | | | |
| 79-80 | | | | | | | |
| | | | | 80-81 | | | |
| | | | | 81-82 | | | |
| | | | 82-83 | | | | |
| | | | | | 83-84 | | |
| | | | | | 84-85 | | |
| | | | | 85-86 | | | |
| | 86-87 | | | | | | |
| | | 87-88 | | | | | |
| | | | | | | | 88-89 |
| | | | | 89-90 | | | |
| | | | | 90-91 | | | |
| | | 91-92 | | | | | |
| | | | | 92-93 | | | |
| | | | | 93-94 | | | |
| | 94-95 | | | | | | |

| | | El Niño | | Neutral | | La Niña | |
|---|---|---|---|---|---|---|---|
| Weak | Moderate | Strong | V. strong | | Weak | Moderate | Strong |
| | | | | | | 95-96 | |
| | | | | 96-97 | | | |
| | | | 97-98 | | | | |
| | | | | | | | 98-99 |
| | | | | | | | 99-00 |
| | | | | | 00-01 | | |
| | | | | 01-02 | | | |
| | 02-03 | | | | | | |
| | | | | 03-04 | | | |
| 04-05 | | | | | | | |
| | | | | | 05-06 | | |
| 06-07 | | | | | | | |
| | | | | | | | 07-08 |
| | | | | | 08-09 | | |
| | 09-10 | | | | | | |
| | | | | | | | 10-11 |
| | | | | | | 11-12 | |
| | | | | 12-13 | | | |
| | | | | 13-14 | | | |
| 14-15 | | | | | | | |
| | | | 15-16 | | | | |
| | | | | | 16-17 | | |

Events in Table 14.1 are defined as 5 consecutive overlapping 3-month periods at or above the +0.5° anomaly for warm (El Niño) events and at or below the -0.5° anomaly for cold (La Niña events. The threshold is further broken down into *Weak* (with a 0.5 to 0.9 SST anomaly), *Moderate* (1.0 to 1.4), *Strong* (1.5 to 1.9) and *Very Strong* (≥ 2.0) events.

"Late in September of 1982 the sea surface temperature rose 4°C in 24 hours along the seacoast near the town of Paita, Peru. Officials in this peaceful village were immediately on the alert for El Niño – an ocean warming phenomenon associated with reductions in fish, birds and marine mammals. Little did they anticipate the depth and destruction that were to affect their town and the country of Peru. This El Niño event was to be one of the worst to impact South America."

The above quote was the first paragraph of a TOGA brochure that the author had prepared as Director of the International TOGA Project Office[2]. [It was a brochure prepared and distributed in four languages -- and many individuals and organizations helped in that preparation].

The devastation from the event had attracted the interest of many countries: scientists from the four countries for whom the brochures were prepared, and other scientists interested in the international project. Scientists participated in various ways: helping to provide atmosphere/ocean observations with new observing systems, analyzing the new data produced for further understanding of the phenomena of El Niño and La Niña, and providing new predictive models for these events.

The atmosphere's input is the Southern Oscillation Index (SOI) which is the normalized difference in sea level pressure between Tahiti, French Polynesia and Darwin, Australia. That pressure gradient (difference) represents a proxy for the strength of the southeasterly trade winds across the Pacific Ocean. The positive values of the Index imply *strong trade winds* which are associated with La Niña (the Tahiti *surface pressure in the east higher relative to the Darwin pressure* in the west – driving strong winds from east to west). The negative values of the Index implied a reversal in that pressure gradient and a weakening of the trade winds associated with El Niño.

The ocean's role is determined primarily by the depth of the *thermocline*. The *thermocline* is a *region* between the "surface layer" a relatively warm shallow layer of ocean waters which are well mixed with a fairly constant vertical temperature; and between the "deep ocean waters" *below the thermocline* which have a much colder temperatures, also with little change in temperature with depth. The *intermediate temperatures* of the *thermocline decrease rapidly* from the "surface layer" to the "deep layer".

The *thermocline* depth in the eastern Pacific near South America is quite shallow during neutral or La Niña condition allowing nutrients brought to the surface by upwelling processes to be available to the normal marine life in the region. When El Niño occurs the thermocline layer is driven down and there are no longer upwelled nutrients – the fish move elsewhere or die. The catches of several species of fish were severely depressed during 1982-83 El Niño. The anchoveta disappeared. These fishing losses were severe for Peru and Ecuador but less than other economic losses that were to occur.

The loss of the stabilizing influence of the cool air in the eastern Pacific led to dramatic changes in the climate patterns of the entire eastern Pacific.

Rainfall in Ecuador and Peru was intense and persistent until July of 1983. Along the coast of Ecuador the rainfall reached 30 times the normal average. In northern Peru the rainfall reached as high as 140 times the normal average. The widespread flooding took its toll on crops, livestock, roads, bridges, schools, homes, and human life. In Ecuador 40,000 families lost their homes in total or in part. In Peru the number was 50,000.

The climate conditions in the western side of the Pacific and Indian Ocean changed significantly as well. The anomalous westerly winds would continue for nearly a year, into the first week of June, 1983 -- until abruptly ending. The winds reflected the record swing in the Southern Oscillation index.

Surface pressure over the Indian Ocean began to rise in the last half of 1981 and drought conditions began to spread from southern India eastward to Indonesia, Australia, and the Philippines. The five-month running mean surface pressure at Darwin, Australia reached its highest recorded value in mid-1982. Australia experienced its worst drought in 200 years. Immense dust storms, numerous brush fires, and a two-billion dollar loss in agriculture and livestock resulted.

A great many new observing systems, primarily oceanic in nature, were put in place for the TOGA program and many of these are still in place and have improved the current knowledge of these climatic oscillations. A partial list of these observing systems includes: an island and coastal tide gauge network for sea level measurements; drifting buoy arrays to provide mixed layer velocity and SST measurements; the Tropical Atmosphere-Ocean (TAO) array of moored buoys to provide surface wind, SST, upper ocean temperature, and current measurements[3].

The current status of advanced prediction of these El Niño and La Niña events has some good news and some bad. The predictions from models both physical and statistical have been positive for 1- month and 4-month lead times, but basically worthless for the 7-month lead times[4]. The timing has been pretty good, the predicted intensities not so good.

Looking back on Table 14.1 one can see a pretty chaotic picture of events. Excluding the *Weak* El Niño events, the *Moderate to Very Strong* events occur every 3 to 6 years. The only clear fact is that after two consecutive periods of *Neutral* events there has always been an El Niño event – and only one has been *Weak*, the others *Moderate, Strong* or *Very Strong*.

Very good progress resulted from this TOGA program. There is sufficient evidence from those results that a further international effort could lead to further substantial benefits! There remains more than enough economic incentive to pursue this climate research activity *again*. Two of the Very Strong El Niño events were in 1982

-1983 and again in 1997-1998. The 82-83 event cost 2000 lives to perish and the economic impact was estimated at $13 billion. The 97-98 El Niño event was worse and there were 21,000 live lost and the economic toll was $35 billion[5].

This author believes that a broader program with new and improved observational equipment, better satellite support now available, and with sufficient new funding becoming achievable; that a repeat international project like TOGA could and should be implemented.

It will take some time for all the interested parties to band together again and produce a convincing scientific plan for a second phase – call it TOGA-II or perhaps a better name. Scientists should take adequate time to schedule an international program for the 10 year period 2022 to 2031.

There are multiple scientific papers that speak of a cooling that will occur in the 2030 time period – plus or minus 5 years. This is discussed in the last Chapter of this book. Should that cooling materialize, there should be sufficient signs in advance for a *further designated international effort that* could couple with an existing TOGA-II to help decipher the *possible implications of the degree of cooling that might occur.* Thus the two programs would support each other.

There is another climate-change space/time scenario that could also benefit from the TOGA–II activity and this is further research on the Pacific Decadal Oscillation which has both broader space and time scales than the El Niño application.

The Pacific Decadal Oscillation (PDO) represents a long-lived pattern of sea level pressure and temperature changes that resemble the El Niño oscillation is some ways -- these two Pacific Ocean climate change patterns have somewhat similar spatial characteristics However, the PDO time scales is considerable longer – ranging from two distinct time scales of between 15 to 25 years and 50 to 70 years[6].

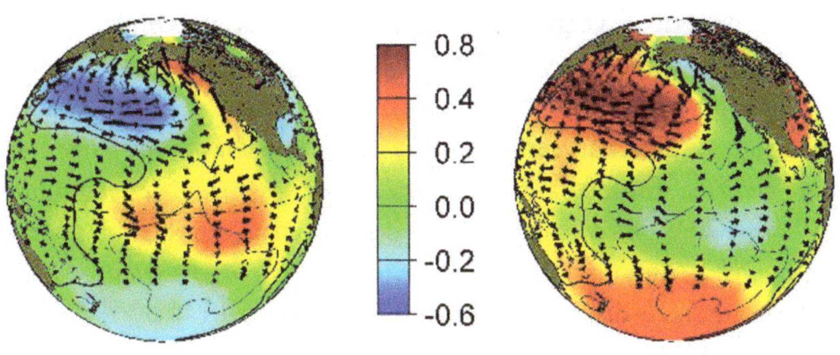

Figure 14.1  PDO Warm phase on left and cold phase on the right

Figure 14.1 indicates the sea surface temperatures (colors), the sea level pressures (contours) and the surface wind stress (arrows) anomaly patterns during the warm and cool phases of the PDO[7]. One can see that in the warm phase, shown on the left of the figure, the *positive* SST anomaly is in the eastern North Pacific along the west coast of North America. There is a large area of *negative* SST anomalies in the western and central North Pacific extending eastward from Japan.

There have been only two full cycles in the 20th century: cool periods in 1890 to 1924 and again in 1947 to 1976; and warm periods in 1925 to 1946 and again in 1977 to 1998. Major changes in the northeast Pacific marine ecosystems have been correlated with the phases of the PDO. Warm eras of the PDO have led to *enhanced* coastal ocean biological productivity in Alaska; while at the same time *limiting* such productivity along the west coast of the United States. The cold periods of the PDO have displayed the exact opposite conditions in the respective regions.

At this time there is no single well established cause for the Pacific Decadal Oscillations – though there are speculations as to perhaps multiple causes. Clearly the coupled atmosphere and oceans are capable of producing such changes. The question is one of how can the atmosphere/ocean coupling produce such long term timing – and the related concern of whether extra-terrestrial forces also contribute. It is possible that three international projects could be on-going at once!

Chapter 12 indicates the Sun's magnetic field, its motion about the solar system barycenter and cosmic rays have influenced *climate-changes*. These solar system *oscillations range from periods of 60 to 350 years and longer*. Scafetta[8] has identified a number of climatic changes over time periods which match those *shorter periods* -- which he feels are *likely* driven by astronomical changes.

Scafetta has listed the PDO mentioned above, the Atlantic Multidecadal Oscillation (AMO), quasi-60 year periodicity in secular monsoon rainfall records from India, proxies of monsoon rainfall from Arabian Sea sediments, and in rainfall over east China as potential examples. He claims that the empirical evidence is there that these climate changes are *likely* driven by astronomical causes.

Though Scafetta believes the causes of these shorter time period climate changes are likely solar system induced, he also agrees that the actual physical mechanisms causing the oscillations have yet to be identified. From what we have learned about the longer term changes discussed in our travels so far, we are perhaps close to an explanation. However, there are several areas that require further research and hopefully the funding agencies will open up the playing field for these efforts.

There is much to be learned about the detailed inner workings of the Sun. NASA plans to expand the monitoring of the Sun. The Parker Solar Probe was launched in August of 2018. This is the first-ever mission to "touch" the Sun. The spacecraft, about the size of a small car, will travel directly into the Sun's atmosphere to about 4 million miles from the surface.

On November 16, 2018 the spacecraft first transmission verified that all sensors were performing as expected. Each of the orbits about the Sun will be petal-shaped – where the spacecraft will skim closely to the Sun, then fly out farther into space to close out the orbit. When the spacecraft travels close to the Sun and has a speed comparable to that of the Sun's rotation – then it can keep pace with that rotation; it will be observing the same region of the Sun over a period of about 10-days – observing the detailed dynamics of the Sun over that time period[9].

During the mission the spacecraft will become closer and closer to the Sun, eventually coming within 4 million miles of the surface of the Sun. On each orbit the same measurements will be taken at different depths within the Sun's atmosphere, the corona. Scientists are hopeful to understand how the coronal gets so hot (~1 million degrees) and how the Sun produces phenomena like the solar wind and solar flares. The mission should last until 2025, if the fuel lasts that long (the fuel is used to twist the spacecraft to keep delicate instruments hidden behind a heat shield).

Other required research is needed to improve the possible link of solar influence on all the various space/time climate scenarios. The cosmic ray background is not likely to change much in our lifetime. However, the degree of cosmic rays entering our atmosphere will primarily depend on the activity of the Sun's magnetic field. The Parker Solar Probe will provide a great deal of new knowledge about the Sun.

A secondary help will be a better global network of detecting the key isotopes produced by cosmic rays, over space and time. Greater evaluation of the isotopes themselves is important, and of course, this will help verify the status of the Sun's magnetic field.

There is another breakthrough, which has not yet occurred. This would be to understand the details *of dark matter*. The results of the Planck space observatory, a satellite launched by the European Space Agency in 2009, confirmed that the fundamental constituents of the universe have been refined to be approximately 68.3% dark energy, 26.8% *dark matter* and 4.9% ordinary matter (visible stars, planets, interplanetary gases). [See Appendix A for further history of the universe and how it evolved after it was formed].

Dark matter does not emit or interact with observable radiation, but is affected by the force of gravity. It is detected by several methods used by astronomers. Does dark matter has any influence on cosmic rays? We suspect not as it has been a fixture of the universe long before

Earth and our solar system was formed. Nevertheless, one hopes to soon remove any possible connection it might have with cosmic rays.

## Chapter 14 Future Energy Related Actions

The fact that the $CO_2$ released from the burning of fossil fuel has no impact on climate-change allows the world to re-think the optimal way to use all fuels to supply the world's energy requirements. It was pointed out in Chapter 13 that: (1) $CO_2$ is essential to life and will be more valuable if the climate cools significantly; (2) $CO_2$ is not toxic and its former contribution to air pollution has been eliminated since 1975 by catalytic converters and by similar equipment to remove toxins and pollution from the world's power producing plants; and (3) the scare of ocean acidification is a misrepresentation of the facts.

The world has been encouraged/pushed to switch to renewable energy and abandon fossil fuel energy regardless of the cost. The world cannot afford that plan! This world effort need not happen now! An estimate was made in 2008 by the International Energy Agency of the cost of reducing global manmade $CO_2$ emissions 50% by 2050 – the estimate was $45 trillion – we have now just saved that amount! Now, rather the governments of the world can carefully analyze the cost/benefits of a combined role of fossil fuel, hydro-power, nuclear energy and renewable energies -- taking into account actual fuel costs, maintenance costs, and unnecessary subsidies.

It is appropriate to consider the history of how the world has come to this important decision point with regard to the energy choices of fossil fuels and the other energy forms.

Coal, oil and the rest of the fossil fuels were formed millions of years ago – before humans existed. The benevolent Creator of our universe foresaw the future energy requirements for mankind. Think about the status of whales on our planet – they were on their way to extinction due to mankind's need for lamp oil just prior to the sudden commercial availability of fossil fuels. While many different species of creation have come and gone, God saved the whales.

Norwegians began hunting whales 4000 years ago. During the Middle Ages and Renaissance whaling became more popular throughout Northern Europe. Every part of the whale was used: meat, skin, organs and blubber – the latter being the source of whale oil – primarily used for oil lamps. Whaling spread to North America, but by the 1700s it was increasing difficult to find whales along the Atlantic coast.

Whaling operations spread to the whale-rich waters of the Arctic and Antarctic. Whaling in America hit its peak in the mid-1800s then quickly died off with the first commercial oil wells and an abundance of coal became available – electricity delivered to the home eliminated the need for oil lamps[11].

Coal was produced by plants and the Sun millions of year's ago[12]. Much of the Earth was covered by huge swamps filled with giant ferns and plants. They died, sank to the bottom, and were later covered by soil and water -- after much time passed the pressure and heat changed the plants into coal. America has one-fourth of the known coal in the world and nearly 300 billion tons of recoverable coal, a 250-year supply if used at the current rate. Many countries have coal – China is the world's largest user with about 5.5 times the amount of USA's coal use, and ~ 60 other countries are listed as using some coal.

Coal is used in power plants producing about half of the electricity used in the USA. The process of converting the energy in coal to electricity involves several processes. The coal is pulverized into a fine powder where it is mixed with hot air and sent to a *furnace* to burn and heat water in a *boiler* to create steam. The steam spins the blades of an engine called a *turbine*. The spinning turbine powers a *generator* that converts mechanical energy into electrical energy (this happens when magnets spin within copper coils in the generator). A final process uses a *condenser* to cool the steam moving through the turbine condensing the steam back into water, where it returns to the boiler and the cycle begins again.

Electricity-generating power plants (both coal and natural gas powered) send out electricity using a *transformer,* changing electricity from low voltage to high – a step giving electricity the power it needs to travel from the power station to it final destination. At this point voltages can be as high as 500,000 volts. Electricity flows along transmission lines to *substation transformers.* These substation transformers reduce the voltage for use in the local areas to be served. Electricity travels from the substation transformers along distribution lines, that can be either above or below the ground, to cities and towns. Transformers again reduce the voltage to ~ 120 to 140 volts for safe use inside businesses and homes. The delivery process is instantaneous – one flips a switch to turn on a light -- and saves a whale.

Coal is not perfect, it contains sulfur, nitrogen, and tiny specs of minerals and dirt. However, the process of *gasification* can remove 99.9% of the sulfur and dirt particles – using lots of heat and water. The total costs of producing electricity from coal are slightly less than producing it from natural gas. This leads us to provide some further detail about the other fossil fuels of oil and natural gas.

The history of petroleum/oil is long and need not be included here but can be seen on the internet[13]. Today about of 90% of the fuel requirements for the various surface vehicles are met by oil. Petroleum represents 40% of the total energy consumption in the USA, but is only responsible for 2% of electricity generation. The value of petroleum as a portable, dense energy source (powering the vast majority of vehicles) and as the base of many industrial chemicals, makes it one of the world's most important commodities.

Commercial jet airplanes require a fuel with *high energy density* and Jet A and Jet A-1 fuels are a mixture of different hydrocarbons produced to a standard international specification.

Shown in Figure 14.2 are conceptual illustrations of various types of oil and gas wells. A vertical well is producing from a conventional oil and gas deposit (right). Also shown are wells producing from unconventional formations: a vertical coalbed methane well (second from right); a horizontal well producing from a shale formation (center); and a well producing from a tight sand formation (left)[14]

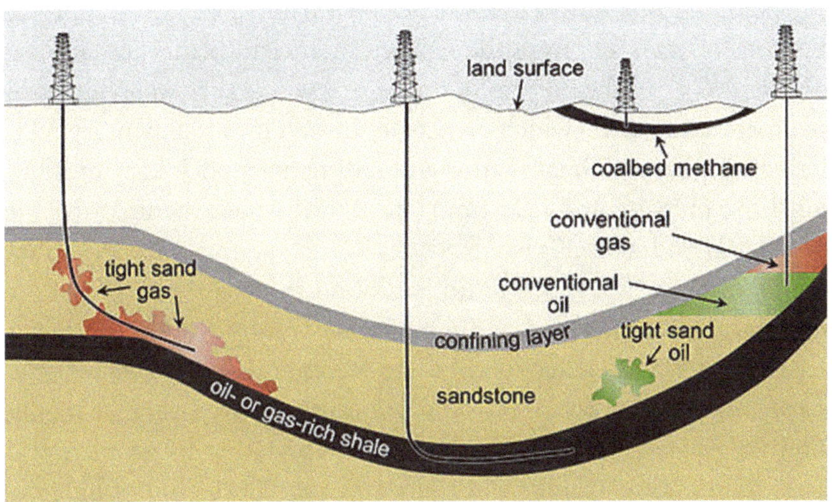


Figure 14.2 A conceptual illustration of various types of oil and gas wells.

Shale oil is an unconventional type of oil produced from oil shale rock fragments. Details on it is extracted can be found in the reference[15]. This process has been around for a long time but has become more popular of late. An even newer method of extraction is that used for "tight oil" also known as *shale-hosted oil or light tight oil abbreviated LTO*. Details on tight oil can be found in the reference[14].

The three top oil producing countries are Saudi Arabia, Russia, and the USA. About 80% of the world's readily accessible reserves are located in the Middle East. A Table at the end of this energy summary reveals the relative uses of the various fuels (fossil, nuclear, hydro-electricity and renewables) of various countries and the world as a whole.

*Natural gas was* found naturally since ancient times – people of Greece, India and Persia discovered natural gas many centuries ago. Natural gas produced from coal was used in England for lighting as early as 1785. Natural gas seeping from cracks in the ground would occasionally be struck by lightning and cause strange fires that mystified the populace. It has been used commercially only in fairly recent times.

Natural gas is a fossil fuel like petroleum and coal as it was formed millions of years ago from the remains of ancient plants and animals died and sank to the bottom of the oceans where they were buried by sediment and sand – which eventually turned into sedimentary rock. Layers of plants, animal matter and sedimentary rock kept building until the pressure and heat from the Earth turned the mixture into petroleum and natural gas.

Natural gas is trapped in underground rocks and was difficult to find in early times. Now it is discovered *onshore* by artificially creating seismic waves. The reflection of these waves are captured by sensitive *geophones* and sent on to a seismic recording facility (truck or mobile site) that records the data for further interpretation by geophysicists and engineers for analysis – as the waves are reflected off of different layers of the Earth.

The original seismic waves were generated by carefully placed small explosions using dynamite. This has since given way to non-explosive seismic technology such as a large piston from special vehicles that impact the Earth creating the waves, or from *offshore* discovery ships that tow an array of hydrophones -- the seismic wave source in this case is a large air gun that sends bursts of compressed air – sending the waves through the ocean and down through the Earth's crust[16].

Effective pipelines began to be built in the 20th century thus allowing natural gas to be used in manufacturing and in boilers to generate electricity. It is the home consumer that has really benefited from the expanded pipeline delivery system with natural gas used for home heating, cooking, and for appliances such as water heaters, clothes dryers and oven ranges. There is a two million-mile delivery system in the USA that has an outstanding safety record.

The use of natural gas is growing steadily throughout the world. Extracting gas from gas-rich shale is becoming less expensive as better methods evolve. There are six nations with *vast deposits* of gas-rich shale: the United States, China, Argentina, Algeria, Canada and Mexico hold an estimated 80 % of documented shale gas deposits – other countries will eventually come along as well. Natural gas is by far the cheapest and cleanest fossil fuel. The building of *liquid natural gas* terminals around the world will continue to expand the use of natural gas – a far better way to produce electricity than by the use of renewable sources of energy.

British Petroleum publishes a statistical review of world energy use each year and their report for 2018 summarizes the growth of energy use in 2017[17]. Natural gas consumption rose by 96 *billion cubic meters (bcm)* or 3% in 2017, the fastest growth since 2010. Global natural gas production increased by 131 bcm, or 4%, almost double the 10-year average growth rate. *Gas trade* expanded by 63 bcm, or 6.2%, with growth in liquid natural gas (LNG) leading the growth in pipeline trade. The increase in gas exports was driven by Australian and USA LNG (up by 17 and 13 bcm respectively), and Russian *pipeline* exports (15 bcm).

Before covering the renewable energy resources, it is quite informative to observe the progress seen in renewable energy over the past few years as reviewed by the Group chief executive of British Petroleum in the introduction to the BP Statistical Review of World Energy 2018 (their 67th consecutive annual report.) The Group chief executive was Bob Dudley.

From 2014 through 2016 saw three consecutive years where there was *little or no growth in carbon emissions from energy consumption*. "This came about through accelerated gains in energy efficiency muting growth in energy demand, and rapid growth in renewable energy combined with successive falls in global coal consumption leading to improvements in the fuel mix."

2017 saw *a reversal as carbon emissions from energy consumption grew* – energy demand picked up and *coal consumption increased for the first time in four years*. Renewable energy growth was strong, but natural gas had the largest growth – primarily from the coal-to-gas switching in China. *Three years with no progress, then a year of reversal!*

The whole world has been pushed to abide by the IPCC and United Nations declarations to reducing the $CO_2$ emissions to avoid a spectrum of national disasters to come. It has not worked. The experiment in molding governments and people has failed.

The price individuals have had to pay for much higher gasoline costs for autos, and for higher electricity costs for their homes has been staggering – not to mention the higher taxes required to support a huge world-wide bureaucracy of government officials and on the payroll scientists demanding more powerful computers for climate models that do not work – this has been the *real global catastrophe* that has produced *nothing!*

A further quote from Group chief executive Bob Dudley:

"The power sector really matters; it absorbs more primary energy than in any other sector. It accounts for over a third of carbon emissions from energy consumption. However, despite the huge policy push encouraging a switch away from coal and the rapid expansion of renewable energy in recent years, *there has been no improvement in the mix of fuels feeding the global power sector over the past 20 years.*"

Astonishingly, the share of coal in 2017 was exactly the same as in 1998. The share of non-fossil fuels was actually lower, as growth in renewables has failed to compensate for the decline in nuclear energy."

Table 14.2 is from British Petroleum and has been reduced to fit on a single page. The use of the various fuels required by the power sector to produce electricity is organized by the fuels: oil, natural gas, coal, nuclear energy, hydro-electric energy and renewable energy; and organized by the countries of the world.

Table 14.2 The primary energy consumption in 2017 by fuels used by the world – see text for units

| Million tons of Oil Equivalent | Oil | Natural gas | Coal | Nuclear energy | Hydro-electric. | Renew-ables | Total |
|---|---|---|---|---|---|---|---|
| USA | 913.3 | 635.8 | 332.1 | 191.7 | 67.1 | 94.8 | 2234.8 |
| Canada | 108.6 | 99.5 | 18.6 | 21.9 | 89.8 | 10.3 | 348.7 |
| Total North America | 1108.6 | 810.7 | 363.8 | 216.1 | 164.1 | 109.5 | 2772.8 |
| Argentina | 31.6 | 41.7 | 1.1 | 1.4 | 9.4 | 0.7 | 85.9 |
| Brazil | 135.6 | 33.0 | 16.5 | 3.6 | 83.6 | 22.2 | 294.4 |
| Total South & Central Amer. | 318.8 | 149.1 | 32.7 | 5.0 | 162.3 | 32.6 | 700.6 |
| France | 79.7 | 38.5 | 9.1 | 90.1 | 11.1 | 9.4 | 237.9 |
| Germany | 119.8 | 77.5 | 71.3 | 17.2 | 4.5 | 44.8 | 335.1 |
| United Kingdom | 76.3 | 67.7 | 9.0 | 15.9 | 1.3 | 21.0 | 191.2 |
| Total Europe | 731.2 | 457.2 | 296.4 | 192.5 | 130.4 | 161.8 | 1969.5 |
| Russian Fed. | 153.0 | 365.2 | 92.3 | 46.0 | 41.5 | 0.3 | 698.3 |
| Ukraine | 10.0 | 25.6 | 24.6 | 19.4 | 2.0 | 0.4 | 82.0 |
| Total CIS | 203.4 | 494.1 | 157.0 | 65.9 | 56.7 | 0.9 | 978.0 |
| Iran | 84.6 | 184.4 | 0.9 | 1.6 | 3.7 | 0.1 | 275.4 |
| Saudi Arabia | 172.4 | 95.8 | 0.1 | 0 | 0 | - | 268.3 |
| Total Middle East | 420.0 | 461.3 | 8.5 | 1.6 | 4.5 | 1.4 | 897.2 |
| Egypt | 39.7 | 48.1 | 0.2 | - | 3.0 | 0.6 | 91.6 |
| South Africa | 28.8 | 3.9 | 82.2 | 3.6 | 0.2 | 2.0 | 120.6 |
| Total Africa | 196.3 | 121.9 | 93.1 | 3.6 | 29.1 | 5.5 | 449.5 |
| China | 608.4 | 206.7 | 1892.6 | 56.2 | 261.5 | 106.7 | 3132.2 |
| South Korea | 129.3 | 42.4 | 86.3 | 33.6 | 0.7 | 3.6 | 295.9 |
| Total Asia Pacific | 1643.4 | 661.8 | 2780.0 | 111.7 | 371.6 | 175.1 | 5743.6 |
| | | | | | | | |
| Total World | 4621.9 | 3156.0 | 3731.5 | 596.4 | 918.6 | 486.8 | 13511.2 |

Table 14.2 provides the BP summary of primary energy consumption of commercially-traded fuels used to generate electricity. [Note the '- 'implies < 0.05.] The units in the Table 14.2 are *millions of tons of oil equivalent (mtoe)*. The original table contains some 66 named-countries, other smaller ones within various regions, and the world totals. Table 14.2 had to be is limited by space, thus includes only 2 or 3 countries with the largest use within a region (with no intent to discriminate in the choice of a country). *All the regions in the original Table* are included in this Table 14.2

Clearly China is the largest user of total energy with 3,132 (mtoe). China uses the most coal and has the largest hydro-electric use. The USA is the second largest user of energy and uses the most oil, the most natural gas, and, (at the moment) is the largest user of renewable energy.

There needs to be a new discussion of the uses of renewable energy – *there is no longer an urgent need to replace fossil fuel!* Various forms of renewable energy (solar, wind and biofuels (like ethanol – already used in gasoline) have a place in the world, but *not for all applications*. Let us now perform a limited but fair comparison of the solar energy source for various applications.

Solar energy has been fantastic for my TI-30Xa hand calculator. I have had it for many years without a problem and it has never been in the Sun – just the diffuse radiation through my home office window has supplied continuous power when needed.

The ability of the Sun to provide direct energy is clear. A solar cell is a composed of a thin semiconductor wafer specifically treated (doped) to form an electric field positive on one side and negative on the opposite side – a pn-junction with a particular bandgap energy, $E_g$. When a photon of light strikes the solar cell, electrons are knocked loose from the atoms in the semiconductor material if the photon energy is $\geq E_g$. [If the reader returns to the first page of Appendix D you will find that the photon contains the energy of the light radiation, and that photon energy is more powerful as one moves further to the left of the spectrum.] Thus a solar cell formed with multiple layers with multiple bandgaps can respond to multiple light wavelengths – capturing more of the Sun's wavelengths. A multijunction cell with a stack of individual single-junction cells with a *descending order of bandgaps* would have the top cell capturing the *high energy photons* and passing the rest of the photons on to be absorbed by lower band-gap cells.

Using solar panels on the roof of a home can lead to potential savings. However, the next step is to have micro inverters under each solar panel so that the DC (direct current) electricity generated by the solar panels can be converted into AC (alternating current) useful in the home. This current can run into a 'net meter' combined with your previous electricity from the local grid and you can use both for all your appliances and needs – before sending any excess power back to your utility company for a refund.

This is a good use of solar electricity for the home, but it may not be good for the community. Why? Because the issue with solar energy is the lack of continuous operation. The electric grid that exists in most countries provides instantaneous energy to one's home by merely flipping a switch as described above within the discussion of power plants providing electricity when powered by coal or natural gas. Whether that electrical grid is powered by coal or natural gas (a cleaner source) in either case the power supplied is continuous.

The potential energy from the Sun to provide sufficient power, at a single site for a sufficiently long period is obviously out of the question – the energy availability is sensitive to 'day versus night,' the weather situation on any given day, winter versus summer, and the potential of very little or no sunlight available over several days in parts of the world. Back-up energy from fossil fuel is required for solar – and for wind energy – the wind must blow.

Simply adding solar or wind energy to an existing electric grid *imposes significant additional costs.* The electric grid must be re-engineered and adapted to operate quite differently in order to function with the erratic nature of solar and wind generated electricity. The existing gas or coal power plants that create the electric grid were designed and optimized for continuous operation – not the randomly intermittent nature of solar and wind energy.

What can one say about wind energy? Denmark has more wind turbines per capita than any nation on Earth – and also the most expensive electricity of any nation[18]. It is intermittent, relatively expensive, has high maintenance, and kills bats and important birds – apparently 2,300 golden eagles have been killed over a 25 year period at Altamont. California. I cannot say anything positive, so I will simply quote Delingpole[18] "Wind Park; Wind Project; Wind Resource Area (WRA) are cozy euphemisms used by the wind industry to make bat-splattering, bird-slicing, landscape-destroying, sleep-ravaging eco-death factories sound nice."

Those who have successfully promoted solar and wind energy have misled everyone into thinking that these are competitive with the traditional energy sources of fossil fuels. However, there are a variety of financial incentives behind the true costs of producing these energy forms.

There are *direct* government subsidies that cover 30 to 50% of the cost of these two renewable energy sources. There are also *indirect* subsidies that allow profitable companies to take advantage of various deals that reduce their tax burden. These are tax equity financing arrangements based upon providing investment capital for solar and wind projects.

Many of the states in America (and in many regions of various countries) the politicians have passed laws requiring that a certain percentage of their electricity come from renewable energies. As a result of these laws, the utility companies are required to pay higher prices and sign long term purchase agreements with solar / wind power developers. The higher prices are simply passed on to the electricity users. These mandated contracts (often with government provided subsidies) makes it possible for the renewable energy companies to finance their investments/projects. A study by the U. of Texas estimated U. S. energy subsidies per megawatt hour (a 1000 kilowatt hour) in 2019 would be $0.5 for coal, $1 - $2 for oil and natural gas, $15 - $75 for wind and $43 - $320 for solar[19].

These energy companies, of course, also support the politicians that created the renewable portfolio laws to begin with. This arrangement is very costly for the electric consumer regardless of his/her financial situation. This is a reverse 'Robin Hood' scenario – not stealing from the rich and giving to the poor – but stealing from the poor and giving to the rich.

Biologically produced alcohols are the most common form of biofuels. World biofuel production increased 3.5% in 2017 well below the 10-year average 11.4%. Brazil and the USA led that increase primarily with ethanol fuels[15] from sugarcane and corn, respectively. The USA no longer requires *ethanol as a security energy backup* with the enormous growth in availability of increases in natural gas and oil via gas and oil shale development. *That land will be far more useful for food production with the expected climate change that is coming.* Delingpole[18] has some choice comments on ethanol. "What ethanol does do very successfully is drive up food and fuel prices, ensure that cars run fewer miles to the gallon, increase water consumption, and help encourage starvation and food riots in the developing world. Very eco-friendly!"

Another form of energy to be briefly mentioned is Nuclear energy The Nuclear energy source of power creation has been scaled down world-wide but there is still plenty of fuel. Uranium (the last and heaviest of the natural elements) has several isotopes, U-235 is the proper fuel in reactors but only is present in 1% of uranium -- compared to U-238 at ~ 99%. However U-238 can be converted into plutonium which can also be used as fuel in a reactor.

Plutonium is the most important of the transuranium elements, all of which follow uranium in the periodic table, and all of which are artificially made[19]. Plutonium-239 is the most important because it readily fissions when bombarded by thermal neutrons. Like U-235, the nuclei of its atoms split into two intermediate-size nuclei (called fission fragments), releasing large amounts of energy and producing more neutrons to sustain a chain reaction.

Plutonium can be produced in huge quantities of tens of thousands of kilograms per year in nuclear reactors. The abundance of Plutonium has made it the material of choice for nuclear weapons.

We have yet to mentioned hydroelectric power production, and this is important for many countries that have the river systems that can produce it. This is very much tied up with water use in general and will be covered in Chapter 15. There are important issues with the use of dams on river systems that are controversial.

Since the late 1970s until the present, there has been 50 years of government propaganda pushing political pressure for renewable energy. Now governments can make rational decisions on the best combination of fuels to use in their power production and in their future automobile manufacturing. The auto industry appears to be headed toward hybrid cars and trucks with both electric and gas power sources – we need not cover that.

We will concentrate on the proper combination of fuels for *primary power production. What individuals do with renewable versus fossil fuel use in their home is their business.* There are a number of important factors to consider for *future power production* and perhaps serious changes to be made in the energy power production field.

Fossil fuels (coal, oil and natural gas) will not run out for another 500-1000 years. The first action is to stop the 'reverse Robin Hood syndrome' which must be broken as it hurts all electricity users.

The second action is a full evaluation of all the economic factors related to power production, and an evaluation of the issues related to maintaining the integrity of the electronic grid (with and without renewable energy resources appended to the electric grid). Including in this evaluation must be the necessary actions to create and fund energy systems required to counter a *blackout* of the grid – whether from a serious solar disturbance or from sabotage by a terrorist group or from another nation. *This is very important and should be a part of the complete evaluation of total energy system performance, fuel costs, cost of maintenance, cost of back-up, and so forth.*

The real cost of primary power production must be determined. All government subsidies, if any, need to be evenly distributed among the various fuels. There are decisions to be made concerning real fuel costs, import vs export, new plant construction costs, operating costs, and concern for the consumers in seeking to eliminate all waste within the grid system.

Those who had believed, and still believe, that *climate-change* is one of the *most important challenges to humanity this century* and that governments have the *moral* responsibility to provide *subsidies* for renewable energy in order to increase the rate at which the world migrates to a lower carbon society – apparently are motivated only by political reasons. They are wrong in every respect and thoughtless beyond comprehension! They apparently do not understand the real world environment nor care to learn anything about it. Let us dissect these views.

First it has been shown by every conceivable point of view that (1) there is no correlation of $CO_2$ impacting climate change in any direction (warm or cold) throughout the historical and the modern observational records; (2) warmer climates (if they occur, and they have in the past, with much warmer temperatures then our current Modern Warming) are not a problem, and societies have thrived in such warm cycles; (3) $CO_2$ is extremely beneficial for humankind and animals and more is needed , not less!

Governments have no *moral* responsibility; they have a *responsibility* to provide for the economic welfare of their citizens who elected them. Fossil fuels have elevated our entire civilization to a plateau of progress, wealth, and standard of living beyond all past societies. The world's populace need to keep electing smart government officials who aspire to make energy costs low and who seek the advice from scientists and engineers on the best way to do so – because it has been demonstrated in the past that low energy costs are tightly coupled to human prosperity.

There needs to be a fresh and serious discussion on the degree that renewables are combined in the power production sequence. Previous decisions were based upon a false premise and not on hard facts. These decisions will impact several generations of Earth's inhabitants over the next several hundred years.

## References: Chapter 14


1.  Null, J., 2018: El Niño and La Niña years and intensities. https://ggweather.com/enso/oni.html.

2.  Fleming, R. J., 1985: Left position as Director of NOAA Office of Climate and Atmospheric Research (later Office of Global Programs) in Suitland MD to become Director of International TOGA Project Office in Boulder, CO. Great move!

3.  McPhaden, M. J., et al., 1998: The Tropical Ocean-Global Atmosphere observing system: A decade of progress. *Journal of Geophysical Research*, **103 (C7)**, 14169-14240.

4.  Barnston, A., 2014: How good have ENSO forecasts been lately? https://www.climate.gov/news-features/blogs/enso/how-good-have-enso-forecasts-been-lately?/

5.  https//library.wmo.itl/pmb-ged/wmo_912_en.pdf

6.  Minobe, S., 1997: A 50 --70 year climatic oscillation over the North Pacific and over North America *Geophysical Research Letters*, **24**, 683 686.

7.  https://researc.jisao.washington.edu/pdo/

8.  Scafetta, N., 2010: Empirical evidence for a celestial origin of the climate oscillations and its implications. *Journal of Atmospheric and Solar-Terrestrial Physics*, **72**, 20 pp.

9.  https://www.nasa.gov/content/goddard/parker-solar-probe

10. https://en.wikipedia.org/wiki/Dark_energy

11. https://www.nationalgeographic.org/news/big-fish-history-whaling/

12. https://www.worldcoal.org/coal/what_coal

13. https://en.wikipedia.org/wiki/History_of_the_petroleum_industry

14. https://en.wikipedia.org/wiki/Shale_oil

15. https://en.wikipedia.org/wiki/Tight_oil

16. http://naturalgas.org/naturalgas/exploration/

17. https://www.bp.com/content/dam/bp/en/corporate/pdf/energy-economics

18. Delingpole, J., 2013: The little green book of eco-fascism. Regency Publishing, Washington D.C., 258 pp.

19. https://www.forbes.com/sites/uhenergy/2018/03/23/renewable-energy-subsidies-yes-or-no/

20. Stwertka, A., 2002: *A guide to the elements*. Oxford University

CHAPTER **15**

# Summary and Concern about
# the Future

The summary of the observational evidence for $CO_2$ causing *climate-change* has been performed for every observing period from geological historical records from 850 million years ago to modern measurements from balloons and satellites. The evidence shows that there was no correlation of $CO_2$ values with temperature either in cold or warm *climate-change* regimes. The *apparent correlation* with the rise of $CO_2$ during the Modern Warming was, in reality, a correlation with the *Sun's magnetic field / cosmic ray connection* with the Modern Warming, as proven by the data and the cool period from 1940 to 1975, and as proven by the radiation calculations of $CO_2$ which were presented in Chapter 11 and summarized below. The misrepresentation of $CO_2$ as the cause of the Modern Warming was apparently due to the timing of the Industrial Revolution and perhaps due to ulterior motives of other individuals.

Detailed radiation calculations with thousands of absorption lines for $CO_2$ reveal that here is no net residual heating due to $CO_2$ by the normal radiation process. Water vapor and $CO_2$ absorb solar energy at the Earth's surface, but then the three processes of convection, latent heat release, and the emittance of diffuse longwave radiation redistribute that heat upward for radiation balance.

One can restate this result in a different manner. The radiative gases of $H_2O$ and $CO_2$ respond to the *daily changes in weather that result from the chaotic baroclinic weather system generated for that day*. The necessary action for these gases is to radiate the required response to blend with the other two atmospheric forces of convection and latent heat release to redistribute that surface heat upward to produce a balanced energy exchange. There is no net accumulation of heat beyond that expected for the seasonal changes throughout the year. This required action drives the atmospheric heat engine, drives the ocean currents, and powers the Earth's life supporting irrigation system.

The calculations were performed with a dry atmosphere. Had clouds been added in some statistical way, the numbers would have been even lower as the clouds tend to cool the atmosphere. Various temperature profiles (both stable and unstable) surrounding the standard atmosphere profile had only a slight impact and did not change the basic conclusions.

It is primarily the effects of the Planck function that cause the rapid decrease in radiation intensity with height. Another reason is that the *longwave radiation is diffuse* which depletes the intensity rapidly over distance. The diffuse nature of the radiation also leads to the fact that the *net back radiation* for a given level (that sent upward at the bottom of a layer, minus that sent downward at the top of a layer) further *slightly* reduces the adsorbed $CO_2$ radiation intensity.

Other so-called "greenhouse gases" including methane (some with larger absorption coefficients, but all with significantly less concentration) have their intensity quickly transferred upward and depleted by the same strong Planck function intensity change that applies to CO2 and H2O.

*$CO_2$ has no impact on climate-change! The value of $CO_2$ is immeasurable – it provides the basis of life on Earth – Chapter 13 summarized those $CO_2$ benefits.*

The cause of these climate transitions in the Earth's interglacial period of the past 11,500 years is the solar magnetic field and cosmic ray connection. When the *solar magnetic field is strong, it acts as a barrier to cosmic rays entering the Earth's atmosphere, clouds decrease and the Earth warms. Conversely when the solar magnetic field* is *weak*, there is no barrier to cosmic rays – *muons*, produced by cosmic rays, greatly increase large areas of low-level clouds, increasing the Earth's albedo and the planet cools.

The above explanation of *climate-change* applies over many different time scales. The timing of the passage of the solar system as it travels through the spiral arms of the Milky Way Galaxy coincides with the formation of all the Ice Ages on Earth.

Those on the pro-side of $CO_2$ causing *climate-change* have only their 'climate models'. This is like 'standing on sinking sand' as these models have yet to provide accurate results. Why are their results so poor – especially in the tropical atmosphere? See Fig 15.1 from Douglass, et al[1].

Predicting *climate-change* via numerical models is a very difficult problem! The dialog here will not delve into all the potential problems for climate modelers, but will cover two important points. The first point is the very poor climate model performance in the upper atmosphere of the tropical region. The Fig 15.1 exhibits observational data and climate model results over trend lines for the *time period of 1979 to 1996* -- this from data published in 2004 (models have shown little improvement since that date, so the problem still remains).

The Fig. 15.1 indicates temperature *trend-lines* [$10^{-3}$ K/decade] versus log pressure (altitude) for different zonal averages[1]. The four frames in the figure are well marked by region: global, tropics, Northern and Southern Hemispheres. The three different climate models (all marked with open symbols and dotted lines) are: the Hadley CM3 computer model, The DOE Parallel Coupled Model (PCM), and the Goddard Institute for Space Studies (GISS) S12000 model.

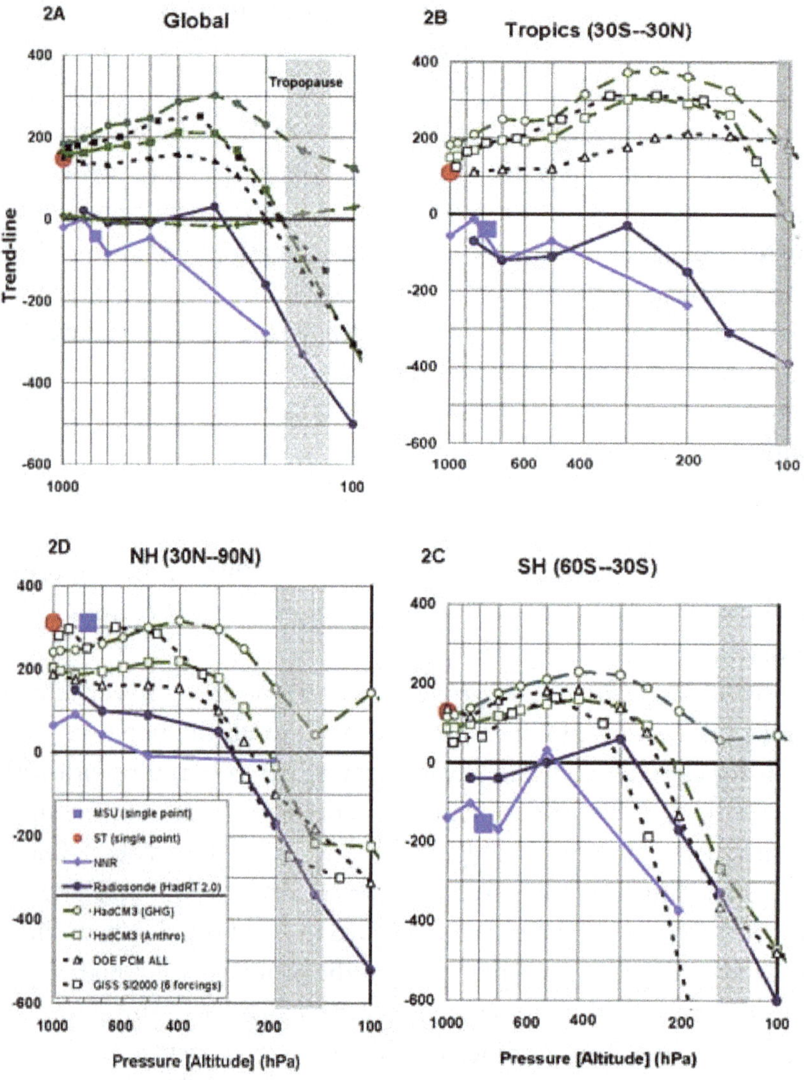


Fig. 15.1  Temperature trend-line information from models and various data sets

The observational data are indicated by the filled symbols and the solid lines; these are: Microwave Sounding Unit (MSU) data (a single fixed point), Surface temperature (ST) from Hadley, (a single fixed point), NCEP/NCAR reanalysis (NNR) a retroactive analysis using data from several different sources over a 50 year period, and Radio-sondes HadRT2.0 which was original radiosonde data giving temperature anomalies at standard pressures, then *trend-lines were computed* from this data by Brown et al.[2].

There are differences in all the sectors of Fig.15.1, *but here we focus on the larges difference is in the tropical region* – all the model trends are *positive* and the observational trends are *negative* – there is a large difference!

There has been no additional heating in the tropical mid-troposphere. But the climate models, especially when they display the impact of a doubling of $CO_2$ always indicate a much warmer tropical troposphere. Our calculations indicate the *initial surface radiation intensity* due to $CO_2$ absorption *decreases rapidly with height, becoming virtually transparent at 16 km in the tropics.*

We know from IPCC reports that they have urged a formula for the modelers to use in the past (from their 2007 Sensitivity Analysis) that expresses their expected delta forcing due to the increased doubling of $CO_2$ with the formula below (this is the exact formula first used by Arrhenius in 1896).

$$\Delta F\ (CO_2) = 5.35 \ln (C/ C_0) = 5.35 \ln (2) = 5.35 \times 0.6931$$
$$= 3.71 \text{ W m}^{-2}.$$

Just how this delta forcing is exactly incorporated into the radiation codes in the climate models is unclear and has not been found in the literature by this author. It appears that this approach alone (or a similar one) would imply that the heating is "baked into the cake" so to speak – perhaps one of the reasons for the poor correlation with reality.

An article in *Science* (27 July 2018) from two climate modelers indicates that there has been 25 years of a lack in improvement in reducing the model errors and the detail of the radiation calculations have not been forthcoming as some would like to see. There appears to be a reluctance to divulge the details of these radiation calculations. Have they been performed accurately as specified in the various radiation text books?

The second point with climate models is that they have a strong positive feedback (perhaps too strong) from increased water vapor, but little negative feedback from increased low level clouds. There is a reason for the positive feedback as increased atmospheric temperature allows the atmosphere to hold more water. However, there is further justification to believe how this can be too strong in the models – as shown in the following discussion.

A paper by Stephens and Ellis[3] discusses the results from a study of 21 coupled climate models which were run as part of an IPCC evaluation. The models were run with an increase of $CO_2$ of 1% per year for 70 years at which point the $CO_2$ would have been doubled from its original concentration. The average results from the 21 models was that the model-predicted *water vapor increases per degree of warming* were more than 3 times the respective rate of increase of precipitation.

The reasons put forth for this result may have had merit, but there is another possible reason that was not mentioned in the paper. There is a particular method for *one of the nonlinear moist physics parametrization tools* that may have been used in these models. This parameterization was probably well-known and shared by the modeling community. Moist physics was important to this author's work in aviation, and this was studied while working at the University Corporation for Atmospheric research (UCAR).

There is an equation in the models: $\partial E / \partial t = + C (E_S - E) +$ [other small positive term]) where E is vapor pressure, $E_S$ is saturation vapor pressure, and C is a complex expression, *small in value, but is always positive.* This vapor pressure increase is at the expense of *cloud water* (the time rate of change of *cloud water* has the right hand side with a *negative* sign instead of a *positive* sign) – depending on the *temperature encountered by the grid point in question.* [There are other terms in the cloud water equation, but the above term will dominate.]

Thus, the direction of the change depends on the current values at each of the 3-dimensional grid points in the model of E, $E_S$ and temperature T. From the numbers below one can see as T increases linearly by 10 degrees C, the saturation vapor pressure $E_S$ *increases in a strongly nonlinear way*!

| T (in °C) | -60 | -50 | -40 | -30 | -20 | -10 | 0 | +10 | +20 | +30 |
|---|---|---|---|---|---|---|---|---|---|---|
| $E_S$ (Pascals) | 1.8 | 6.1 | 18.3 | 50.0 | 124 | 285 | 609 | 1224 | 2330 | 4229 |

If a climate model has a time step of 20 minutes, the number of times each grid point is evaluated is given by (3/hour) (24 hours/ day) (365 days/ year) (70 years) = 1,839,600 times over the climate run of 70 years. Given that the values of T and E are fairly evenly distributed about their mean values of $T_M$ and $E_M$, those T values greater than $T_M$ will have systematically larger $E_S$ values and the term + C ($E_S$ – E) is systematically biased positive – increasing water vapor at the expense of cloud water. With this many encounters over the length of a climate run, it is readily conceivable that this could lead to a *water vapor increase per degree of warming* that was 3 times greater than the precipitation increase per degree of warming.

Most climate modelers agree that the *water vapor increase* raises the warming purportedly due to $CO_2$ by a factor of approximately two. The climate models have other problems that will not be discussed in detail here.

## THE FUTURE

The opinions of three scientists already referenced (Landscheidt, Sharp, and McCracken) have proposed that an end to the current

Modern Warming will occur in the next couple of decades with the next solar minimum. In 2015 and in 2016 there were three additional papers in each year that expressed the same opinion of a new solar minimum in the next few decades.

There were 7 new papers in 2017 calling for a significant cooling – with some providing specific dates of cooling in the 2030-2040 time period, the 2025-2050 period, and the 2020-2053 time period. Already by mid-2018, there had been four other presentations on this expected cooling – all are solar related and some indicate the Sun's motion about the SSB as a condition of concern. It has been reported that 92 new scientific papers in 2018 have linked solar forcing to climate[4].

The predictions mentioned above on the dates of expected cooling are *estimates*, and must be, because as indicated in Chapter 12, there are three factors involved in the solar impact: (1) the solar dynamo (where there is good , but not perfect, knowledge of cause and effect), (2) the motion of the Sun about the SSB (where the past cycles have been identified quite well), and (3) the density of cosmic rays which is assumed to be fairly uniform when the Sun is not within a spiral arm of the Milky Way Galaxy.

The *center* of the 70 - year Maunder Minimum, the strongest and coldest period within the Little Ice Age (1300-1850), occurred in 1680 of the Little Ice Age. If one adds the 350 year cycle suggested by McCracken to 1680, one arrives at the center of the next solar minimum in 2030 – just 11 years away from the time of this writing in early 2019. Should we be alarmed, perhaps not; but should we be concerned – absolutely, and we must plan for such a contingency.

One must also remember that the Earth's atmosphere has tremendous variability – and is chaotic as shown is Chapter 4. The intense variability occurs is all climate regimes – warm or cold.

Another less scientific approach of simple pattern recognition of previous sunspot cycles has been discussed on the internet for some time. An illustration found on the internet and referred to in various references is presented as Fig. 15.2. There is concern about the next

Solar Minimum which falls in the time window of Fig 15.2. Sunspots #20 through #23 match the pattern of #1 through #4. Moreover, the maximum for cycle #24 was nearly one year later than projected and is *considerably lower in the maximum ISN (International Sunspot Number) that was projected.*

The ISN number has been on a steady decline since the maximum of Cycle #21 occurred in December of 1979. The sequence of these maximum ISN numbers has been: #21 (ISN = 232.9) > #22 (ISN = 212.5) > #23 (ISN = 180.3) > #24 (ISN = 116.4) – a systematic decrease with time.

The ISN number for Cycle #24 is the 3rd lowest since these numbers have been accurately recorded since 1755. The lowest of that era was 81.2 in 1816 (in the middle of the Dalton Minimum, part of the Little Ice Age); the 2nd lowest was 101.6 in 1906.

The original projection based upon the Butterfly diagrams as appear in Fig 12.10, was for the ISN of #24 to be *larger than #23 and its maximum was to occur in 2011 or 2012.* The actual maximum is *significantly lower than #23 and* occurred in April of 2014. Figure 15.3 indicates the status of this cycle as of November, 2018.

*The longer length of Cycle 24 and its lower amplitude* have suggested in the past that Cycle #25 will be even lower. This does suggest that a significant solar Minimum is forming. There is uncertainty in the *magnitude* of the minimum and in the *exact timing of its temperature impact.*

Recall that the Little Ice Age had four Solar Minimums over its 550 year appearance from 1300 to 1850. Plimer[5] more precisely calls the first one the Wolf Solar Minimum 1280 – 1340; he suggests that the Little Ice Age (LIA) started in 1303 and *that only 23 years separated The Medieval Warming ending in 1280 from the start of the LIA in 1303.* The other three Minima were the Spörer Minimum (1450 – 1540), the Maunder Minimum (1645-1715) and the Dalton (1795-1825).

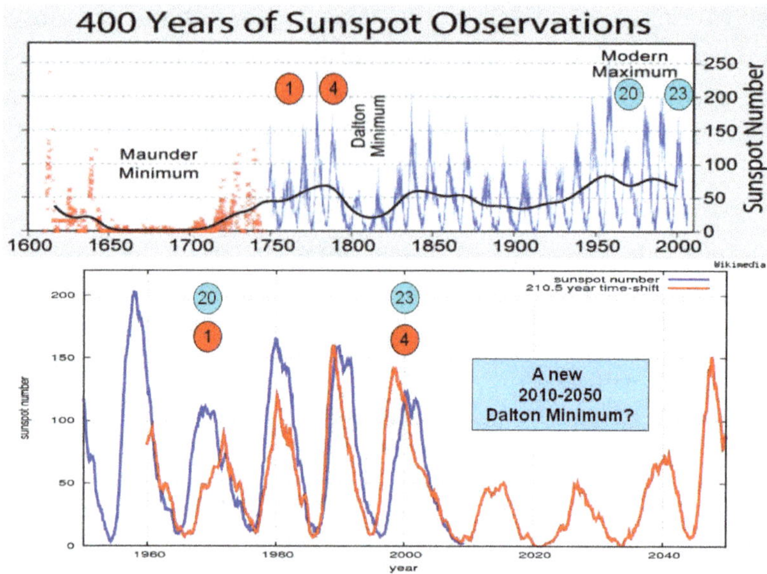


Fig 15.2  Sunspot comparison of current trend with the past

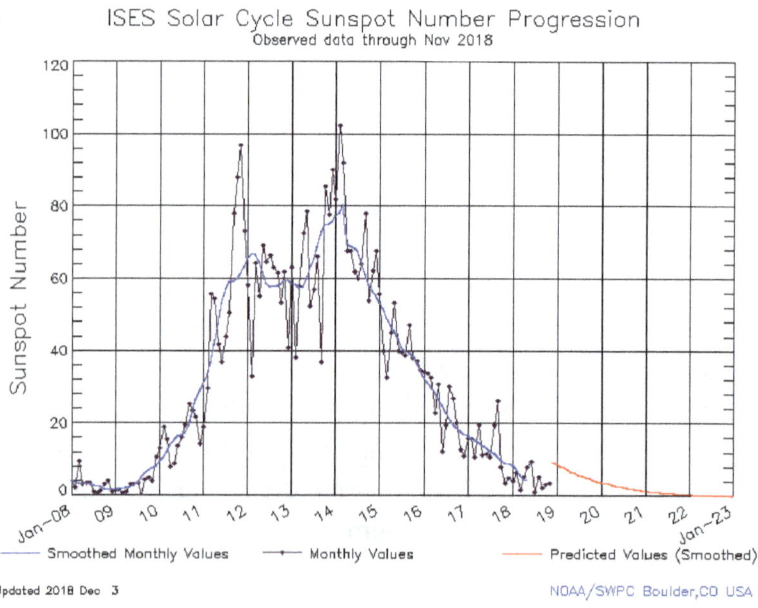


Figure 15.3  Sunspot cycle with emphasis on the latest Cycle #24

An update to Sunspot Cycle #24 is shown in Fig 15.3 from the NOAA Space Weather Prediction Center (SWPC[6]). This indicates that Cycle #24 has been declining more quickly than forecast. The smoothed, predicted sunspot number for April to May, 2018 was about 15, however actual values have been lower. The SWPC panel's prediction for Cycle #24 was a peak of 90 occurring in May of 2013 – the actual result was a peak value of 82 in April of 2014 – nearly a year later – because of its extended length, this also suggests a lower amplitude for Cycle #25.

There is another aspect of the Sun that sheds further light on its dynamic status. This is the solar impact on the Earth's thermosphere. What is that?

Above the troposphere is the stratosphere, the stable thermal layer directly heated from solar radiation absorbed primarily by ozone -- this layer extends from 6.2 to 31 miles above the Earth's surface. Above the stratosphere is the mesosphere where the temperature *decreases* throughout the layer – the coldest temperature in the Earth's atmosphere is about -90°C (-130°F) – this extends from 31 to 53 miles above the surface. Above the mesosphere is the thermosphere where solar activity strongly influences the temperature which increases with height.

The thermosphere is where the aurora occur (the Northern and Southern Lights) where various ions collide with molecules at high latitudes and the extra energy leads to the emission of photons of light – the thermosphere extends from 53 to approximately 621 miles above the surface. When the Sun is very active at the peak of the sunspot cycle, X-rays and ultraviolet radiation from the Sun heat the thermosphere – raising its height to ~ 620 miles above the surface. On the other hand, when the Sun is at the low point of the solar cycle, the solar radiation is less intense and the top of the thermosphere lowers to ~ 310 miles[7].

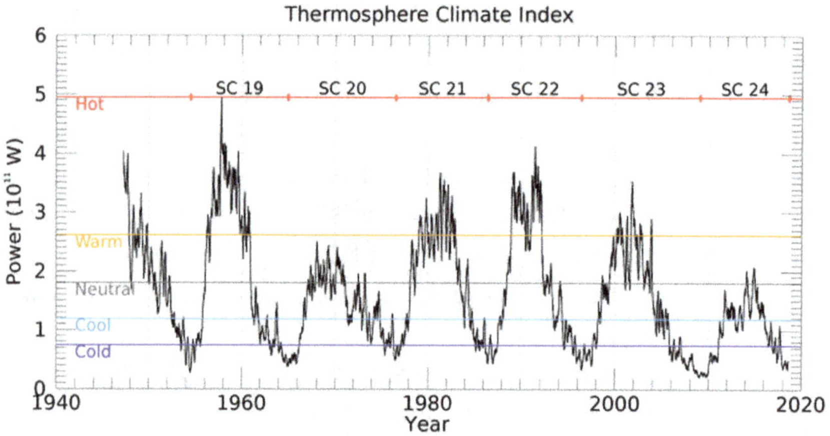


Fig. 15.4  Historical record of the Thermosphere Climate Index

Mlynczac, et al.[7] have a paper that suggests that satellite measurements of the infrared emissions from the thermosphere are nearing a record low values and may set a record by the end of 2018. Fig. 15.4 indicates this infrared power received from nitric oxide (NO) and other molecules. Mlynczac claims that they are currently measuring energy (Watts) at a rate that is 10 times smaller than seen in peak periods. This is not a new *climate-change* source, but a reflection of the intensity of the solar dynamo changes that we have been discussing. It is an excellent reflection!

A Grand Minima appearing in the 2030 ± 5 year time frame – has been predicted by several scientists. The degree of cooling is not certain – some say a Dalton type Minimum and some say a Maunder type Minimum which brought severe impacts to society in the Little Ice Age.

A few facts about the Maunder Minimum (which lasted for the 70 years from 1645 to 1715) need to be addressed. Selections below, progressing forward in time over the above period, are from weather records from various countries. These are primarily from Europe, but similar results were seen in China, India and the USA. The references are found in Marusek[8] unless otherwise noted.

There was famine in Scotland and in the northern part of England in 1649 and 1650. There then followed a plague in England and Ireland. In 1650 and 1651 there was a famine throughout Ireland.

The winter of 1664-1665 was very severe in France. Belgium had very severe frosts and heavy snowfalls. The winter in Poland was most severe. Most of the wines froze, people lost their limbs due to severe frostbite, and others froze to death.

Much of the Northern Hemisphere experience a "year without a summer" in 1675[9]. The winter of 1676-1677 was extremely cold in northern France. The Seine River at Paris was frozen for 35 consecutive days.

The winter of 1683-1684 lasted for 13 weeks in England. The Thames River in London was frozen to a depth of 11 inches and booths and shops were erected on the ice. Many trees, plants and birds were destroyed by the extreme cold. Ice formed for a time between Dover, England and Calais, France; Ice was 27 inches thick in the harbor of Copenhagen, Denmark.

Northern Ireland had a famine in 1689 – some inhabitants ate rats, tallow and hides[9]. An extreme snowstorm occurred in Scotland in 1690 which lasted 13 days and nights – during that time 90% of the sheep froze to death and many shepherds lost their lives.

The winter of 1691-1692 was especially severe in Russia and Germany where many cattle and people froze to death. Due to the intense cold, wolves came into Vienna, Austria attacking people.

A general picture of the causes of the famine problems in Europe is summarized by Plimer[5] in the following paragraph: "Land abatement, crop failure and soil losses were catastrophic because 90% of the population were subsistence farm families who needed enough grain to see them through the winter and enough spare grain to sow for the following year's crop. Both the quantity and quality of harvests were vital for survival. Grain rotted in the fields and sometimes couldn't be planted at all. Crop failure led to famine, famine led to disease and death. Famine led to a breakdown in society and even cannibalism. Gangs of desperately hungry peasants roamed the countryside searching for food."

The above details do not paint a pretty picture for the future *if* the next cool period is as bad as that of the Maunder Minimum. However, the Dalton Minimum (which followed) was not as bad, and lasted only 30 years (1795 to 1825) -- details below are from Marusek[8] unless otherwise indicated.

There was a famine in Europe in 1816. The eruption of Tambora in Indonesia on 10 April 1816 was another factor in diminishing the Sun's intensity in a large part of the world.

The winter of 1815-1816 was known as the year without a summer. Three long cold periods had extreme effects on Canada and the New England region of the USA. The first period in June killed most of the crops. The second period in July killed replanted crops. The third period in August killed corn, beans, potatoes and grape vines.

The state of Connecticut had their coldest temperatures ever recorded and 1816 was the coldest year on record in the USA[5].

The cold years of 1816 and 1817 created a food crises and widespread unrest in Europe and especially in France. This accelerated immigration to the United States and many American farmers migrated south to warmer latitudes. In the UK the average temperature was 2°C colder and it rained or snowed almost every day. There were crop failures in Bengal in 1816 that triggered an outbreak of cholera – this spread from Bengal and was the world's first cholera pandemic.

With a pending cold period and the many potential problems that could surface, one must look at the world population that exits today and see how it is increasing. There are population problems ahead, but fortunately the long term trend is finally changing.

Throughout history there have been three trends in world population growth[10]. The first period of 'pre-modernity' was a very long term period of slow population growth. The second period beginning with 'modernity' in 1800 had an *increasing growth rate over time* that reached its highest value in 1962. This was attributed to rising standards of living and improving health standards. This current third period is underway.

The population growth *rate is falling* and is expected to continue to fall (but the growth rate is still positive) -- leading to an end of population growth by the end of this century. This is good news for future generations, but it is *still bad news for the coming cool down* where the world population is expected to be ~ 8 billion in 2024 – and it was only ~ 1 billion in 1800 when there was famine during the Dalton Minimum of the Little Ice Age.

A cool down in time is certain and the solar dynamo activity appears to be already decreasing. The Earth's surface temperatures have remained relatively constant since 2000. Since the world's oceans hold ~ 22 times the heat held by the atmosphere, the pause in the atmospheric warming may be due to the ocean's delayed effect in warming the atmosphere.

The degree of the upcoming cooling is *clearly uncertain* – it could be something *far less than* the Dalton Minimum, *equal to that* of the Dalton Minimum, *or equivalent to the Maunder Minimum* -- clearly plans must be formulated for this range of contingencies.

Governments need to begin making plans soon. If there were famines in the LIA with just one billion people on the planet, how will the world cope with famines with greater than 8 billion people expected on the planet in 2030? Just how the world's governments will react to famines from crop failures around the planet is a question one doesn't want to think about. Nevertheless, contingency plans must be prepared, *and the content of this book would be in-complete without some discussion of potential required actions.*

The worst case scenario could lead to global misery and death on a large scale. On the positive side, humanity now has far greater knowledge and technology available compared to the capabilities of our ancestors of earlier times. Some of these are assets are listed below.

*$CO_2$ does not cause climate change!* A complete reversal of all current world government policies of limiting $CO_2$ production must be abandoned. Research from 100's of biological papers indicate the

value of increased $CO_2$ for food production – even in cooler conditions.

A policy of procuring and implementing *cost-effective* composite energy systems (made up of fossil fuel and non-fossil fuel energy sources – (which work together both day and night, and in all seasons of the year; and with phony subsidies removed) will make a huge difference in efficiency and perhaps save tax funds for the contingency plans that must be formed.

Greater fuel efficiency in the winter months may save many lives for those under-privileged with few financial resources. Fuel costs for heating must be reduced for everyone.

This cost-effective composite energy system could make energy and water irrigation available to every person on Earth – especially helping the continent of Africa which has long been restricted in fully pursuing fossil fuel production by unfortunate UN policies.

Positive biological effects of $CO_2$ enrichment have been shown in many hundreds of peer-reviewed papers. Greater plant productivity with higher $CO_2$ concentration, even with cooler growing conditions, have been shown. These attributes will be extremely helpful in the coming cool period.

Concern about the future *climate-change* requires that we continue to study it in earnest – but in a different direction! This effort must focus on the natural galactic *climate-change* that is upon us. Billions of dollars have been wasted chasing the wrong problem. It required a great deal of pride for scientists to think that they could *control* the Earth's climate. We must forget that error and now simply strive to understand other forms of climate change that affect humanity.

The various scales of *climate-change* involve a mixture of scientific disciplines and galactic forces. There are real climate issues to solve now! We cannot change what will happen, but we can prepare for the change with plans that cover all possibilities.

There is a very important control *problem required* of humanity that must be improved upon immediately! This task was detailed in

Chapter 9. The Earth's irrigation system has been supplied with three of the four important components: the power source, the water source, and the distribution and drainage system – but humanity has been left with the most difficult task of *management and control of the Earth's irrigation system.* Our past progress has been poor.

Past cold periods have experienced severe floods as significant snow and ice build-ups occurred in the winter months – then the spring thaw brought unprecedented flooding. Humankind has made improvements in flood control, but the extent of the flooding during the Little Ice Age has probably not been matched since. Countries will have to re-evaluate flood control on rivers and creeks to see if further improvements might be needed – not just for minimizing losses due to floods, but also for capturing more fresh water wherever and whenever possible.

This issue of fresh water for food production and controls of potential flooding is extremely complex and important. Some decisions will have to be made in advance. This requires some further explanation of the various issues involved.

There is going to be major social debate (involving economic and political fallout) coming with the expected *climate-change* cool-down involved with the Earth's hydrological cycle – this involves the delicate management of the fresh water required for food production, drinking water and the importance of flood control. Past cold periods have had both droughts and floods.

The cooler atmosphere holds less water, hence the probability of droughts is higher. On the other hand there will still be floods: (1) the atmosphere is still chaotic (perhaps more so in cooler periods[11]) so there will be convective activity on all space/time scales and the occasional massive rains, and (2) the much colder winters will have more snow and ice – this will lead to spring thaws and larger floods than those experienced over the past 170 year of the Modern Warming.

The debates will occur in many countries, but a brief review of a very important issue in the USA will provide the background of the concern. Dams helped the United States meet the needs of a growing nation serving as the solution to irrigate arid agricultural land, supplying fresh water, and controlling floods. By the mid 1960's dams were being constructed at a fast pace with little concern about the impacts on the environment or outdoor recreation.

Public opposition to the dam's impact on fish and wildlife led to the enactment in 1968 of the Wild and Scenic Rivers Act in an effort to restore some balance to the nation's rivers after years of massive federal dam projects[12]. The good news is that over the past 50 years there have been nearly 12,000 river miles of damaged river beds that have been cleaned up. There is still concern about many rivers in the western part of the US that are affected by dams.

Dams have very important functions and also some unwanted virtues. The first function of a dam is to store water for meeting changing water requirements due to a fluctuating river flow or due to varying water demand for agriculture, industry and individual homes. The second purpose is to raise the level of water upstream to increase the 'hydraulic head' – the difference in height of the surface reservoir and the river downstream. The creation of storage and head allows dams to generate electricity (formerly producing 20% of the world's electricity, but now a lower percent due to more power plants using natural gas for the production). Other positive assets of large dams are providing improved river navigation, reservoir fisheries and leisure activities.

There are three types of dams that are chosen for the various rivers and watersheds – *embankment, gravity* and *arch*. Earth and rock *embankments* represent 80% of the large dams. They are built across broad valleys. *Gravity* dams are thick straight walls of concrete built across relatively narrow valleys with firm bedrock. The *arched* dams are concrete and confined to *very narrow* canyons with strong rock walls and represent only 4% of the large dams. The *arch dam* can be built with only a fraction of the concrete needed for a *gravity* dam of similar height[12].

Severe weather events over many years have weakened *embankment* dams. The earthen dam above the town of Johnstown, Pennsylvania once held back the largest reservoir in the United States, but it collapsed in 1882 and swept 2,200 people to their deaths. There are now 3,500 such dams that are considered unsafe or deficient in the United States. Very large floods are rare, but they do occur. When floods do cause a dam to break or overtop a dam, the very large amounts of water, released all at once, takes lives, destroys businesses and homes, and cause tremendous economic damage.

The aging of dams is one of their unwanted virtues, but there are others. Here are other reasons to *avoid having dams*. Low–lying land on either side of a river's banks make up a river's floodplains; and are naturally a part of the river – they nurture life on the plains and provide a natural protection against floods. Small, regular floods that inundate riverside floodplains are essential to a river's health, and provide a wide variety of benefits to wildlife, fish and people.

Rivers deposit sediment and nutrients in floodplains – making them productive areas for growing crops. Finally, during floods the water over a large floodplain can replenish groundwater supplies – capturing flood water during wet years is one of the best ways to provide adequate groundwater during droughts.

On paper this elimination of the many smaller dams makes sense. However, the emotional impact of the issue between the protections of lives and property from flooding versus the necessity of maximizing food production via expanded floodplains is staggering.

There is a relatively long lead time in *removing* or *repairing* a dam (regardless of the cost of that decision -- a major issue itself). Those pushing for the new theory of transitions toward *adaptive management of water resources* in the face of climate and global change[13] will be *fully challenged* – far more than they had anticipated. Now, rather than preparing for a slowly varying climate warming, they have the potential of a quickly occurring cold climate – with the degree of that *coldness quite uncertain.*

There is another important aspect of the hydrological cycle that must be addressed. There are 66 major *aquifers* in the world and probably many smaller ones that have yet to be identified. These are defined as an underground layer of water-bearing permeable rock, rock fractures, or a combination of rock, gravel, sand and silt. Many of these have been heavily accessed for their water and there is concern about their replenishing.

Clearly, maximum use must be made of these aquifers in the future. They represent a major contribution to the surface water and ground water that is available – all sources will be required for the needed water for the various uses that have already been discussed.

One of the largest aquifer in the world is the Ogallala aquifer in the United States (there are larger ones in Australia, South America and Africa) which covers 450,000 square kilometers and stores as much water as one of the Great Lakes (Lake Huron). The water was deposited about 10 million years ago from runoff from the Rocky Mountains. The aquifer underlies eight different states, but is primarily under the state of Nebraska.

Even this large body of water is constantly being recharged with surface water, but the input has not matched the drawdown from irrigation. Water conservation is applied here and must be on all the world's aquifers. Additional efforts to find and implement other smaller aquifers will be extremely useful for additional fresh water supplies before the cool period occurs.

There is another important activity that will help in some regions of the world which is *actively* returning excess surface rain water into groundwater. Because of the seasonal monsoon, India has a problem of too little rain, then too much rain with farms left waterlogged for months unable to grow further crops. In the dry season a salty layer is created from the standing water which has dissolved the many minerals in the soil.

The British Press has described the work of two gentleman that have created a simple but effective system which allows two crops to be grown in a single year. There are now more than 3500 bhungroo – a Gujarati word meaning "straw" installed across key areas. This is a pipe (4-6 inches in diameter) which is placed in low-lying areas where water logging is a problem. During monsoons the excess water drains down the pipe, is then *filtered*, and continues down below the ground to natural aquifers to be used later[14].

A similar method can be used across all river floodplains to improve groundwater transport to natural aquifers before it evaporates. This will also help the problem seen in the Little Ice Age where standing floodwater became breeding areas for disease carrying mosquitos (any of a variety of insects that suck the blood on animals and humans.)

Another system that mankind has implemented copies natures desalination way of providing fresh rain water – by large scale evaporation from the oceans using the Sun's energy (2500 Joules per gram) as described in Chapter 9. There has been progress with *man's system of desalination*.

It should be noted that nearly 40 % of the world population live within 100 km of oceans or seas. This is an important area for progress to be achieved now and even more so in the future. The cost to build desalination plants has declined which bodes well for the acceptance and growth of desalination. In the 1960's the cost was approximately $10 per cubic meter, today it has decreased by a factor of 10 to approximately $1 per cubic meter ($3.79 per 100 gallons).

The Little Ice Age began in 1303, and 15 years later the Bubonic Plague struck the Chinese Gobi Desert. It eventually killed 35 million Asians, then spread westward and killed ~ one third of the European population. It was known as the Black Death but was a combination of the bubonic plague, septicemic plague and the pneumonic plague[5].

This was and is a major health concern, but the world's medical capabilities may be sufficient to hold these plagues in check. It is *malnutrition* that is the key issue, therefore the ability *to properly feed upwards of 8 billion people* is the concern – *how to create, store and distribute food*. Unfortunately, malnutrition is still a great problem for many today!

Medical procedures are now far better to deal with the various plagues that have sprung up during the past cold periods. A string of major and minor famines, leads to malnutrition and a weakened immune system. Increased floods produced new swamplands that become mosquito breeding grounds that spread tropical diseases like *malaria* that swept through Europe[8]. The discussion of actions related to food are in Chapter 16.

## References: Chapter 15


1. Douglass, D. H., B. D. Pearson and S.F. Singer, (2004): Altitude dependence of atmospheric temperature trends: climate models versus observations. *Geophys. Res. Lett.*, **31**, Li320.

2. Brown, S. D. Parker and D. Sexton, (2000): Differential changes in observed surface and atmospheric temperature since 1979. *Hadley Cent. Tech. Note HCTN 12.* Hadley Cent. For Clim. Pre- dict. And Res. Devon, U.K. Stephens, G. L. and T. D.

3. Ellis, (2008): Controls of global-mean precipitation increases in global warming GCM experiments. *Journal of Climate,* **21**, 6141-6155. http://notrickszone. com/2018/12/27/92-new-papers-link-solar- forcings-to-climate-s

4. ome-predict-solar-induced global cooling by 2030. Plimer, I., **(2009):** *Heaven and Earth – global warming: the miss- ing science.* Quartet Books, Limited, London, 503 pp. https://www.swpc.

5. noaa.gov/products/solar-cycle-progression Mlynczak, M. G., L. A. Hunt, J. M. Russell and B. T. Marshall, (2018):

6. Thermosphere climate indexes: percentile ranges and ad- jectival

7. descriptors. *J. of Atmos. and Solar-Terrestrial* Physics. J.jastp. 218.04.004. Marusek, J.A., 2010: A chronological listing of early weather events, Impact, http://www. breadandbutterscience. com/Weather. pdf.

8.

9.  Parker, G., 2013: Global crisis (war, climate change and catastrophe in the Seventeenth Century). *Yale University Press*, **77**, New Haven, 55-73.

10. Roser, M and E. Ortiz-Ospina, 2013, updated 2017: World population growth

11. Fleming, R. J., 2015: *Analysis of the SDE/Monte Carlo approach in studying nonlinear systems.* Lambert Academic Publishing, Saarbrücken, Germany, 136 pp.

12. de Silva, S. H. C., https//www.internationalrivers.org/dams-what-they-are-what-they-do.

13. Pahl-Wostl, C., 2006: Transitions towards adaptive management of water facing climate and global change. *Springer Science + Business Media B.V.*

14. https://www.bbc.com/news/business_44071895

# CHAPTER 16

# Tipping Points, Turning Points and Future Actions

*The story alluded to in Chapter 1 should end here with all the proofs in hand.* However, there are a few USA politicians and perhaps others in the world that still claim that unless the world does something soon, the world will end in 10 years due to global warming. These are fanatical views. There will always be some on the left whose political views will not allow them to yield to scientific facts – even as the predicted catastrophic events fail to occur.

These false warnings have been labeled as *tipping points*. They have been created by some environmentalists who feel that the Earth is fragile, and needs to be protected – even to the extent, if necessary, of doing harm to humankind. Some feel that the climate system is being pushed beyond its limits, beyond some imaginary *tipping point* from *which there is no return*.

A summary of the progression of these imaginary tipping points or arbitrary deadlines over time is provided in the book of Morano[1]. Only a few of these are mentioned here. In 1982 the UN expressed the warning that the world would face an ecological disaster as final as nuclear war *within a couple of decades* unless the government took immediate action. In 2014 the UN declared that nations needed to take significant action *in the next 15 years* to cut carbon emissions in order to avoid the worst effects of global warming. Al Gore presented his own tipping point in 2006 and again in 2008 that we have *less than 10 years* to make dramatic changes or we would lose our ability to recover from the environmental crises.

A final example captured by Morano and briefly summarized here is the experience of Prince Charles, heir to the British Throne. In March of 2009, the Prince announced that we had *less than 100 months* to change our behavior concerning fossil fuel or risk catastrophic climate change. That date came and went in the summer of 2017. However, two years earlier the Prince had extended the deadline to 2050 – 35 years later, at which time Prince Charles would be 102 years old! *There will be no tipping point – beyond which the Earth cannot recover!* $CO_2$ is not the cause of climate change and the *climate will continue to cycle between warm and cold as it has for all time!*

There is an important *turning point* rather than a tipping point that is more relevant to society now. The *turning point* would be when the current *Modern Warming* ends and natural cooling begins to occur -- as in all past climate change regimes. Earlier, it was pointed out that the date 2030 ±5 could signal a Solar Minimum. However weather is extremely chaotic and is the climate! We do not know now whether this *turning point* has been achieved yet. It may not occur in a given year, but may only be recognized until after the fact as a 5 or 10 year *period of transition.*

It will not be easy to achieve international planning for a cooldown while the UN is still pushing a global warming set of actions — set to cost the nations trillions of dollars. However, the news is encouraging. The world is finally coming around to challenge the United Nations push for global socialism under the veil of catastrophic climate change. A registered letter has been sent and received by the *Secretary General* of the United Nations on October1 of 2019. It was sent on behalf of the *Ambassadors of the European Climate Delegation* which included the European countries of (Belgium, France, Italy, Ireland, Germany, The Netherlands, Norway, Sweden and the UK) and the countries of Australia, English Canada, French Canada, New Zealand and the USA.

The cover letter was signed by delegates from the above countries and the attachment contained the signtures from a network of 500 knowledgeable and experienced scientists and professionals in climate and related fields. The main message was "there is no climate emergency." Some of the sub-points of the letter covered such topics as: the naturally occurring warm and cold periods throughout the history of the Earth's atmosphere; the shortcomings of climate models being used as policy tools; the beneficial nature of $CO_2$; global warming has not increased natural disasters; and there being no cause for panic and alarm, climate policy must respect scientific and economic realities. These topics and more have been discussed in this book.

The main media did not pick up this development, but Michele Stirling of the Friends of Science Society, simply read the contents of the letter in a YouTube presentation. As of noon on January 22, 2020 the number of viewers had reached 695,800 (and counting)[2]. The people of the world are tiring of the debate and seek the truth! When the turning point is identified, serious planning must begin – better yet, the planning should start early!

There will be some warning of the impending cold period from several sources. The Sun is observed daily now from surface systems and by satellite surveillance. Our knowledge of the Sun is immeasurably better than several hundred years ago, and still more powerful satellite systems to come may help remove the remaining mysteries of the Sun.

The date of 2030 was 350 years from the center of the 70-year long Maunder Minimum (the coldest of four periods of the Little Ice Age). There is therefore uncertainty in the timing of the next Solar Minimum. The solar signals are already here and there have been signs of significant cooling in late 2019. Perhaps this solar minimum has already started, however it is too early to be *sure* that the *turning point* has been reached! Knowing when it has been reached is crucial as the planning activities for food production and distribution are extremely important and require significant lead time.

Four countries that produce the most food now are China (number one in production of rice, wheat, onions and cabbage), India (heavy production of food grains – oats, brown rice, barley, spelt and Khorasan wheat), the United States (number one in production of corn, soybeans and total beef output), and Brazil (coffee, sugarcane, corn, soybeans and second in beef output). Countries in the European Union produces vast quantities of food, but a *continent not in the above list* is Africa.

Africa could have the potential of becoming a major food producer -- if capital investment deliberately occurs from other countries on a massive organized scale – this is needed, not only for the African people, but as an additional food source for those 8 billion people.

Here is a suggestion on the order of importance of this investment. First implement an electricity grid for the 600 million Africans currently without basic energy service. The fossil fuel reserves in Africa must be strongly implemented along with solar energy. Even the World Bank, which has long pushed solar energy now agrees that fossil fuels must be used. Africa has coal, oil and natural gas that can be harnessed and combined with a properly developed solar network.

The two things people need most are food and *fresh water;* in Africa the latter is of vital concern. In parallel with the development of the power system required for the electricity grid, is a continent-wide fresh water distribution system and an irrigation system.

Food is a basic need for survival. In an emergency, stockpiling food, even for a week or so, saves having to buy, barter or trade for it. People are familiar with buying food from grocery stores. During a famine these assets will be quickly exhausted.

There are important tips for individual families concerning food storage, proper clothing, backup heating plans, and other tips for survival available from Marusek[3]. However, the food issue will require *considerable planning by governments of the world* with important new ways of doing business over and above what is routinely performed by the current international food distribution system.

Governments must establish a significant numbers of large water-proof grain bins in various regions of several countries. This must be implemented prior to any significant downturn in the temperatures that restrict farm production. Establishing these grain bins must be performed with gamma ray irradiation, and performed in an oxygen excluded environment with low temperatures --to destroy insects and microorganisms, and minimize the loss of natural vitamins[3].

There is a further issue with food produced by the large corporations that must be settled. This is the use of weed control by the massive agribusinesses. There is concern about the possible cancer causing attributes associated with using certain weed control procedures. This must be fully sorted out by all parties involved before massive food distribution systems are used to solve famine problems brought on by a possible *severe* cold period.

Famines affect domesticated livestock due to a severe shortage of livestock feed. Herds should be reduced and only breeding stocks maintained. After lean years, governments should restock strategic reserves of grains and other foods -- using these times to replenish the herds of livestock[3].

There is another very important activity for governments to provide serious attention. This is organizing the populace to contribute to every need that arises. Just one example would be to use every un-employed person – be they able-bodied men and women (not raising young children), *as an employed person* in *some capacity to assist those in need*; e.g., this could be working part-time in a government sponsored breadline. No one should be idle!

A further extremely important government action required is the clear establishment of a proper system of law and order that negates the kind of activity seen in the Maunder Minimum of roving gangs across the country side seeking food.

The author wishes to close this discussion with something positive about the future. This won't be easy as humanity has always thrived during the warm periods and struggled during the cold times. Civilization has seen 170 years of slow steady warmth since 1850. No living person today has experienced a cold period beyond a winter season. One cannot get excited by a cool down! One might wish for the next Solar Grand Minimum not be so "grand" but nature is in control. *The nature of weather and climate has always been chaotic.*

Nations will need to come together with willing hearts to prepare and commit to viable contingency plans for the feeding and watering of the world population during the impending cold period – however mild or however strong it becomes. A World-Wide Summit may be needed -- bringing together skilled people with positive attitudes about making the world a better place to live.

The most important story for the twenty-first century may be the World-Wide planning process described above. The positive outcome from such an endeavor may save many lives and leave the world in a far better place for all its inhabitants. This successful process could inspire people and demonstrate the value of helping one another – this may not lead to 'world peace', but maybe just a safer world to live in!

You can help promote the proper planning. Contact your friends, neighbors and local politicians about the looming need to prepare!

## References: Chapter 16


1.  Morano, M., 2018: *The politically incorrect guide to climate change.* Regnery Publishing, Washington D.C., 413 pp.
2.  Michele Stirling: https://youtu.be/GpVBH-Hy5Ow
3.  Marusek, J.A., 2010: A chronological listing of early weather events, Impact, http://www.breadandbutterscience.com/Weather.pdf.

# Appendix A

There is only one universe, but with several mysteries some solved, some still un-solved.

Fortunately, we keep learning more about this magnificent creation! The nuggets that have been mined these past few decades have been extremely exciting. This Appendix will provide a brief background of key astronomical results. A comprehensive explanation of these findings is beyond the scope of this book, and beyond the scope of its author – however my interest has grown over the past 40 years and I have read several books on the subject and have compiled a few things that are worth sharing. These are remarkable and amazing in their own right, but some also pertain to *climate-change* on Earth.

One needs to understand something about particle physics to explain the early universe.

*Particle Physics Summary: A brief review of some of the necessary particles*

Particles form two groups: those with mass -- (protons, neutrons, electrons). Those that mediate forces between matter (bosons: photons, gluons, other bosons) – which are without mass.

*Fermions: Particles that make up matter: Quarks and Leptons:*

Quarks: 6 types: Up    Charm  Top (with electric charge + 2/3)
                 Down  Strange  Bottom (with electric charge -1/3)

Baryons: Quark triplets: Protons (two *Up* quarks, one *Down* quark) charge = 1

Neutrons (one *Up* quark, two *Down* quarks) charge = 0
Hyperons (12 different quark triplets: heavier, unstable with short lives)

Mesons: Formed from 1 quark and its anti-quark; about 2/3 size of proton unstable, short life

| Bosons: | Force | Spin | Electric charge | Mass | Observed? |
|---|---|---|---|---|---|
| Graviton | Gravitational | 2 | 0 | 0 | not yet |
| Photon | Elect-magnetic | 1 | 0 | 0 | yes |
| Gluon | Strong nuclear | 1 | 0 | 0 | indirectly |
| $W^+$ | Weak Nuclear | 1 | +1 | 80 GeV | yes |
| $W^-$ | " " | 1 | -1 | 80 GeV | yes |
| $Z^0$ | " " | 1 | 0 | 91 GeV | yes |
| Higgs | " " | 0 | 0 | 125 GeV | yes |

The gravitational force, though weak between individual particles, has a major effect between two large bodies -- like Earth and the Sun – the Earth orbits the Sun.

The electromagnetic force (radiation) is a repulsive force between two (+) charges and an attractive force between a proton (+) and an electron (-). This force dominates on the small scale and is $1 \times 10^{37}$ stronger than the gravitational force. If a photon collides with an atom and moves the electron further away, the energy of the photon is absorbed.

The strong force (quantum chromodynamics) is about 100 times as strong as the electromagnetic force, $10^6$ times stronger than the weak force and $10^{39}$ times stronger than the gravitational force.

The weak nuclear force (responsible for radioactivity); the vector bosons ($W^+$, $W^-$ and $Z^0$) are massive which accounts for their effect being only over a short range – these three are formed from a Higgs particle which has large mass as seen above. An example of a typical radioactive decay is: $N^0 \rightarrow P^+ + e^- + V^*_e$; where

$N$ = a Neutron $\rightarrow$ a proton + an electron + an electron anti-neutrino

Events of the Big Bang happened very quickly, and the background Cosmic Background temperature was initially extremely hot, then cooled quickly as estimated and displayed in the Table shown below. Some key events may not be included simply to restrict the size of the Table below. The current cosmic microwave background (CMB) radiation temperature after 13.8 billion years since the Big Bang is down to 2.726° Kelvin.

I put together a slide presentation a few years ago on this subject and my summary of events was based upon a variety of books and sources I had gathered in the past. The sequence of events presented here is from that presentation. Details may vary from input from other sources. My use of terms may reflect a laymen's understanding of official astronomical nomenclature.

The sequence of events that occurred in the formation of the universe can be found repeated in many text books (which can deviate in certain details). References are not allowed in an Appendix by this publisher, but check the text for key references. One should not quote my Appendix for any authenticity as there are disagreements on details, and in some cases on major issues – e. g. string theory and details related to inflation.

I have listed my favorite set of books that I have purchased on the subject of the universe at the end of this Appendix. Many excellent books have been written about the evolution of the universe. I apologize for any references that may not get back to the original source, or that may differ from a different source – without knowing which is correct.

The Table A1 lists various events numbered in chronological order since the start of the Big Bang. There then follows a verbal description of each period explaining things as I understand them – probably in extremely simple laymen's terms – you should get the main idea of each event.

Table A1  A chronology of important events after the Big Bang

| # | Time after The Big Bang | Cosmic Background Temperature (°K) | Verbal description of the event in as few words as possible. Each event will be summarized in a separate text later |
|---|---|---|---|
| 1 | $10^{-43}$ secs | $10^{32}$ | Planck epic: four fundamental forces are one force |
| 2 | $10^{-42}$ secs $10^{-36}$ secs | $10^{28}$ | Grand Unification Theory (GUT) - gravity force separates |
| 3 | $10^{-36}$ secs $10^{-12}$ secs | $10^{25}$ | Inflation and Electroweak epic- strong force has separated |
| 4 | $10^{-12}$ secs $10^{-6}$ secs | ~ 1 x $10^{15}$ | Electroweak epic ends and Quark epic begins |
| 5 | $10^{-6}$ secs 1 sec | ~ 1 x $10^{12}$ | Hadron (quarks and mesons) epic to quark confinement |
| 6 | 1 -3 Minutes | ~ 5 x $10^{11}$ | Lepton and anti-leptons dominate until near annihilation |
| 7 | 3- 20 minutes | ~ 2 x $10^{10}$ | Big bang nucleosynthesis: protons and neutrons combine via fusion |
| 8 | 56,000 years | ~ 11 x $10^{3}$ | Energy density of matter > energy density of radiation |
| 9 | 380,000 Years | 3 x $10^{3}$ | Atoms (nucleus and electrons) start to form; light breaks free |
| 10 | 1 billion years | ~ 16 | Life cycle of stars produce the heavier elements that support life |
| 11 | 2-10 billion years | ~ 9.9 – 3.4 | Gravitational attraction pulls large volumes of matter into galaxies |
| 12 | 9.4 – 13.8 billion yrs. | 2.726 | Today our current Solar and Earth support system is perfect for life |

*Time period # 1: The Planck epic: four fundamental forces are one force. Time after Big Bang = $10^{-43}$ seconds. Background temperature = $10^{32}$ K.*

The estimated birth of the universe is 13.8 billion years ago – plus or minus 21 million years. The temperature is so high in this 'Planck' period ( the Planck scale is the scale beyond which current physical theories do not have predictive value) where the four major forces – gravitation, the strong nuclear force, the weak nuclear force and electromagnetism were one fundamental force.

'String theory' had been proposed as a way to unite gravity and quantum mechanics in 1994 but it failed to live up to that promise. There were 11 dimensions of space-time (our current view of four involving space/time and seven further so-call 'compact space dimensions'. In the initial universe it was proposed that all the dimensions are curled up in a 'superball'.

As the universe expands these dimensions begin to uncurl. At $10^{-43}$ seconds after creation, seven of these dimensions stop uncurling, the rest become our observed universe. The theory is that these extra dimensions remain curled up at every location in the universe. The cross section of the curl is only $10^{-34}$ meters or $10^{18}$ times less than the radius of an electron. No instrument today can resolve such small measurements. String theory may have other mathematical uses.

*Time period # 2: The Grand Unification epic where the gravity force separates. Time after Big Bang is $10^{-42}$ to $10^{-36}$ seconds. Background temperature is $10^{28}$ K.*

As the universe expands and cools, it crosses transition temperatures at which forces separate from each other. At this temperature the gravitational force separates from the other three forces forming the so called grand unified epoch.

The gravitational force between the Earth and the Sun is described as an exchange of gravitons between the particles that make up these two bodies. The force-carrying particles (those without mass) do not obey the exclusion principle so there is no limit to the number that can be exchanged. Since the graviton has no mass its force is long range. Gravitons have yet to be detected in gravity waves.

*Time period # 3: Inflation and the Electroweak Epic: Time after the Big Bang $10^{-36}$ to $10^{-12}$ seconds. Background temperature = $10^{25}$ K.*

The temperature has cooled enough that the strong force has separated from the electroweak force. Scientists suggest an extraordinary inflation period of the universe expanding – to explain unsolved mysteries, One is the "horizon' problem: if a light signal is sent to us from a place farther away from us than light would have time to travel, that place is *beyond our horizon.*

The horizon problem arose with the discovery of the cosmic background radiation being so homogeneous in all directions in space. Astronomers finding galaxies 10 billion light years away in two opposite directions implies that light (information) traveled 20 billion years but the universe is only 13.8 billion years old, so there must have been an extraordinary inflation period.

The inflation period was suggested by Alan Guth to occur when the strong electroweak force separated into the strong nuclear force and the electroweak force due to symmetry breaking (symmetry breaking is the process by which a physical system in a symmetrical state breaks up in an asymmetrical state). The two forces decoupled as a result of cosmic cooling and lost their symmetry and the universe became more disordered. In Guth's model of inflation it lasted only a small fraction of a second (from $10^{-36}$ to $10^{-32}$) seconds and space expanded beyond belief.

*Time period # 4: Electroweak Epic ends/Quark Epic begins: Time after Big Bang is $10^{-12}$ to $10^{-6}$ secs. Background temperature ~ $1 \times 10^{15}$ K.*

The big bang produced energy all squeezed into a very small volume – this was all that was initially created – it marked the beginning of time, space and matter. Einstein's formula of energy to mass

$E = m\ c^2$ (energy = mass times speed of light squared) applied as long as radiant energy exceeds a threshold, that energy can spontaneously change into a particle of matter (mass). At $10^{-6}$ seconds the forces had taken their present form, but the temperature is still too high for quarks to bind together to form hadrons (baryons [protons and neutrons] and mesons.

When $T > 10^{15}$ the expanding universe is filled with radiation creating pairs of particles and antiparticles; these pairs were annihilating back into radiation at a high rate. Quarks and antiquarks were created from radiation and annihilated back into radiation at a high rate also.

With further cooling the radiation was less able to able to create quark-antiquark pairs. As they 'froze' out of the radiation background, a greater number of quarks than antiquarks were left over. This was called quark confinement – matter created.

*Time period # 5: Hadron Epic to quark confinement: time after big bang = $10^{-6}$ to 1 s. $T = 10^{12}$.*

Previously, neutrons and protons were rapidly changing into each other through the emission and absorption of neutrinos. . A neutron *alone will decay into a proton, an electron, and an electron antineutrino. A proton alone is stable*! However, striking a proton with an electron antineutrino *with high energy*, results in a neutron and positron (an antielectron). Therefore, neutrons change into protons by themselves, but the reverse requires extra energy from some kind of collision.

At $10^{-6}$ seconds the universe is hot and dense, there were so many electrons and antineutrinos that equal numbers of protons and neutrons were changing. At 1 second the universe had cooled, the neutrinos/antineutrinos decoupled from the rest of matter and radiation; protons were no longer being changed into neutrons, but neutrons were still changing – the result was ~ 7 times more protons then neutrons in the universe.

Hydrogen nucleus requires 1 proton and 0 neutrons; helium nucleus requires 2 protons and 2 neutrons. As a consequence of the excess of protons, there should be more hydrogen than helium today. There is: 75% of the mass of the universe is hydrogen and 24% is helium – validating this description of the early universe.

*Time Period #6: Lepton Epic – leptons and anti-leptons dominate: Time after big bang = 1 to 3 minutes. Background temperature = $5 \times 10^{11}$ K.*

Leptons (electron, muon, tau, and their associated neutrinos) and anti-leptons dominate the mass of the universe. Prior to 4 seconds after the big bang, there were huge numbers of electrons and their antiparticle, the positron (the same mass but positively charged). At earlier times the photons of the radiation field had so much energy that they can convert spontaneously into an electron-positron pair, via Einstein's equation $E = mc^2$. The numbers of electrons, positrons, and photons were about the same, and all were in thermal equilibrium together.

Once the radiation cooled below $10^{10}$ degrees (at 10 seconds after), the photons no longer have enough energy to make electron-positron pairs, so most of these start to annihilate. One part in a billion of the electrons were left after the electron-positron holocaust. There were an equal number of protons (with positive charge) so the universe has a net electric charge of zero. The number of photons per baryon (proton + neutron) is about a billion.

*Time Period # 7: Big bang nucleosynthesis: protons and neutrons combine: Time after big bang = 3 to 20 minutes. Background temperature ~ $2 \times 10^{10}$ K.*

At this point in time the average temperature of the expanding universe is low enough for neutrons and protons to combine together to make nuclei of the lighter elements such as hydrogen, helium, and lithium – in the process called nuclear fusion (nucleosynthesis).

Neutrons and protons attract each other at very short distances. The strong nuclear force that holds them together is "confined" and cancels out at larger distances. In order to form nuclei, neutrons and protons need to spend time in close proximity to each other – this can't happen if the temperature is too high – as they are then moving too fast to spend time near one another.

The majority of the neutrons are found in the abundance of helium. Free neutrons combine with protons to form deuterium (1 proton & 1 neutron). Deuterium rapidly fuses to helium-4 (2 protons & 2 neutrons), and small amounts of lithium. Nucleosynthesis only lasts for about 17 minutes, since the density and temperature of the universe has fallen to the point that fusion cannot continue.

*Time Period # 8: Energy density of matter > energy density of radiation. Time after big bang = 56,000 years. Background radiation ~ 11,000 K.*

As the universe cools more and more matter is being created by the high energy radiation. Through this expansion *matter loses less energy than does the radiation.*

Eventually the energy density of matter (mostly in newly-formed nuclei) becomes larger than the energy density of radiation (mostly massless particles like photons). Matter then dominates in how the universe expands from this era on. The photon energies are still so powerful that that they continually smash electrons free from their atomic orbits.

At the end of this process, photons scatter much more with each other than they do with matter and the exchange between matter and radiation becomes less efficient.

The density of initially *high-density regions of dark matter has been increasing since matter dominated the universe.* Whatever the nature of dark matter, it interacts only weakly with normal matter and radiation.

Dark matter clumps first from the large scale structure seen in the universe – then at later times *normal matter is drawn by gravity into regions of highest density, eventually forming galaxies and galaxy clusters – this explains why dark matter is found outside and surrounding the visible galaxies.*

*Time Period # 9: Atoms form. Light breaks free! Time after big bang = 380,000 years T = 3000*

Just prior to this time there are plenty of protons and other light nuclei in the universe, and there are plenty of electrons; but until now the universe has been too hot and dense for the electrons to be captured by the nucleus without being driven out of orbit by collisions with other particles.

When the temperature cooled to the point where the average speed of an average electron isn't high enough to escape capture by a proton, then atoms start to form. The first atoms were hydrogen, helium, and lithium. This process is called *recombination.*

At the end of recombination, most of the protons of the universe are bound up in neutral atoms. Therefore, the photons mean free path becomes effectively infinite and the photons can travel freely – *the universe has become transparent – light has broken free!* This event is referred to by astronomers as *decoupling.*

The photons present at this time of decoupling are the same photons that we see in the cosmic microwave background (CMB) radiation today – but today they are further cooled by the expansion of the universe since that time.

*Time Period # 10: Lifecycle of stars produces heavier elements: Time after big bang = 1 billion years. Background temperature ~ 16 K.*

Radiation has cooled and decoupled from the matter; almost all the electrons are bound up in the lightest atoms, the gravitational forces become important.

Small fluctuations in the matter density and gravitational field begin to grow and coalesce. Hydrogen gas is pulled together by gravity until the force causes the gas to collapse and ignite through hydrogen fusion to form the first stars.

During the fusion reaction hydrogen atoms combine together to form helium atoms and photons of light are emitted. The star no longer collapses; a state of equilibrium is reached between the inward gravitational collapse of material and the outward pressure caused by the energy given off during the nuclear fusion reactions.

When a star's supply of hydrogen has already fused into the heavier elements, the star's primary energy source is gone. The outer layers of the star implode, then in rebound, explode into supernovas – spewing the stellar debris into space. The elements of life are now present in the universe as star dust.

Full atoms form (nucleus + electrons); different elements depend upon the numbers of neutrons, protons, and electrons. (See Appendix B on elementary chemistry). Nucleosynthesis sets the stage for the formation of atoms, then molecules, then subsequently the formation stars and galaxies.

*Time Period # 11: Formation of galaxies and our Milky Way galaxy. Time after the big bang = 2 to 10 billion years. Background temperature ~ 9.9 to 3.4 K.*

All the elements in the universe today come from the inside of stars. The process of making and injecting them into the universe takes place over a time scale that is the lifetime of a star – from 2 to 10 billion years. New stars are born from old ones.

The more massive a star, the shorter its lifetime – the hotter it needs to be to balance its gravitational attraction – and the hotter, the faster it burns up its fuel. Stars have periodic explosions (novae), but a massive star ends its life in a *supernova.*

Stars ending in a supernova brighten by a factor of $10^7$ for a few days, before becoming either a white dwarf, a neutron star or as a black hole.

*Large volumes of matter collapse to form galaxies; small galaxies merge to form larger ones, and gravitational attraction pulls galaxies toward each other to form groups, clusters, and superclusters.*

Our galaxy, the Milky Way (began forming ~ 8.8 billion years ago) has a diameter of about 100,000 light years and is part of the part of the Virgo supercluster. There are ~ 100 galaxy groups and clusters located within its diameter. The Virgo supercluster is one of an estimated $10^{11}$ galaxies, each with an estimated $10^{11}$ stars – giving an estimated $10^{22}$ stars in the universe.

*Time Period # 12: Formation of Solar System and conditions for life: Time after the big bang = 9.4 to 13.8 billion years. Background temperature = 2.73 K.*

Solar system forms about 4.6 billion years ago. A molecular cloud of hydrogen and traces of other elements began to collapse, forming a large sphere in the center (our Sun and a surrounding disk). The accretion disk would coalesce into a multitude of smaller objects that would become planets, asteroids, and comets.

Our Sun is the right size to consume hydrogen and produce energy at a rate that provides the time and conditions for life to form. Our orbit through space (distance of 150 million km from the Sun) departs from a true circle by only 3%. Were it as elliptical as the orbit of Mars, we would alternate between baking when closer to the Sun and freezing when distant.

Earth contains just enough internal radioactivity to maintain its iron core in a molten state. This produces the magnetic umbrella that only *partially* deflects the solar wind, solar flares, etc., (see Chapter 12 for more details).

Earth's gravity is strong enough to hold the needed gases of our atmosphere, but weak enough to allow lighter noxious gases to escape into space. All this is balanced at just the correct distance from the Sun so that our biosphere is warm enough to maintain water in the liquid, life-supporting, state, but not so warm that it evaporates into space.

The ultimate fate of the universe appears to be determined: clusters of galaxies are receding (moving away) from us – the universe is expanding, in fact the expansion is accelerating.

The effects of gravitation among all the objects of the universe should slow the expansion, this has happened in the past. However, as the universe expands its volume, the density (mass per unit volume) becomes less. It has recently been determined that in the first ~ 7 billion years of the universe, gravity had the upper hand in slowly the effects of dark energy, but in the last ~ 7 billion years the dark energy has surpassed the effects of gravity.

*The author's books suggested for interesting reading about the universe:*

Aczel, A.D., 1999: God's Equation (Einstein, Relativity, and the Expanding Universe)

Goldsmith, D., 2000: The runaway universe (the race to find the future of the universe)

Hawking, S. W., 1988: A Brief History of Time (from the big bang to black holes)

Ross, H., 2008: Why the Universe is the Way it is

Ross, H., 2018: the Creation and the Cosmos (How the Latest Scientific Discoveries reveal God)

Rowan-Robinson, M., 1999: The Nine Numbers of the Cosmos

Weinberg, S., 1993: Dreams of a Final Theory (The Search for the Ultimate Laws of nature

# Appendix B

This Appendix represents a simple introduction to the subject of chemistry. Chemistry deals with matter, its composition and the changes which matter will undergo. Every chemical change involves an energy change as well. Energy is defined as the capacity to do work. Some chemical reactions are promoted not for the products they produce but for the energy produced.

Society recognizes and uses many forms of energy and are constantly converting one form to another. *Kinetic* energy is the energy of a moving object has because it is in motion. The wind, a stream of moving water have kinetic energy. *Potential* energy is energy due to position. An auto on a hilltop, an object held above a floor have potential energy. There are other forms of energy, such as light, sound, electrical energy, chemical energy and heat.

A typical example of energy conversion is the process that produces electric light. For example the solar energy of the Sun evaporates water from the oceans and this water acquires potential energy as it rises to form clouds. When it falls as rain with some flowing to a waterfall where it loses potential energy and acquires kinetic energy. When it goes over a waterfall there is a conversion of potential energy to kinetic as it is used to turn an electric generator. A portion of this energy is converted to electricity (with a loss of heat), Then as the electricity reaches a lamp, it is converted to light (again with a loss of heat).

There is no loss of energy in the above conversion – heat is energy, the lowest form of energy. This illustrates the Law of Conservation of energy – energy cannot be created or destroyed but can be changed in form. Chemistry deals with homogeneous matter (that which appears the same all the way through) which is composed of *elements* or solutions (compounds) – a solution is composed of two or more *elements* – mixed so completely that it appears homogeneous to the eye. *Elements* cannot be simplified by ordinary chemical methods.

One must begin discussion of chemistry with the understanding of the atom. The nucleus of an atom consists of a number of neutrons (mass = $1.67 \times 10^{-27}$ kilograms (kg) with no electric charge), protons (with a mass of 99.86% mass of the neutron and a positive charge of +1) surrounded by electrons (with a mass of 0.054% of the neutron and a negative charge of -1).

Only a few subatomic particles need be mentioned. *Fermions* are particles that make up *matter – quarks* and *leptons.* There are six *quarks* (Up/Down. Charm/Strange, and Top/Bottom). The electric charge on the top quark is 2/3 and on the bottom quark it is – 1/3. The proton is made up of two up quarks and one down quark; thus its charge is (4/3 – 1/3 = 3/3 = 1). The neutron consists of one up quark and two down quarks; thus its charge is (2/3 -2/3 = 0).

*Leptons* include charged leptons and neutral leptons (neutrinos). Only the electron is stable, the others are unstable with short lifetimes. A summary follows:

Electron (mass = 0.5 MeV/c$^2$: charge = $1.602 \times 10^{-19}$ coulombs: charge = - 1

Muon (mass = 105.7 MeV/c$^2$: charge = -1 [The muon is discussed in Appendix E]

Tau (mass = 1777 MeV/c$^2$: charge = -1

Every element's atom has a nucleus surrounded by shells of electrons. The different 'shells' of electrons are illustrated in Figure B2 with the nucleus of an atom in the center with its surrounding shells (orbits) of electrons. Nature's rule for the number of electrons in a shell (N) is 2 N$^2$.

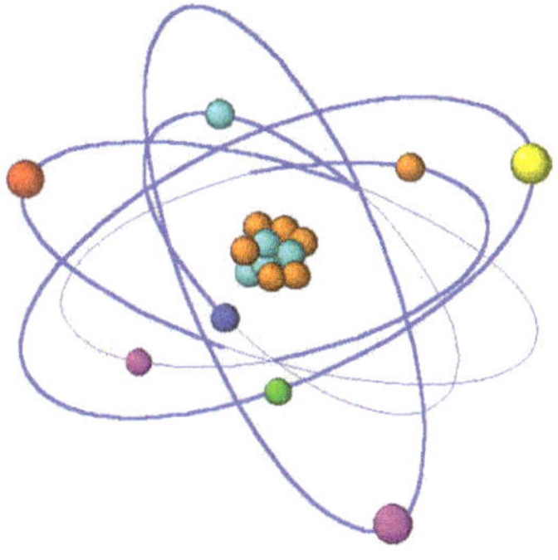

Figure B1  Atom with electrons circling

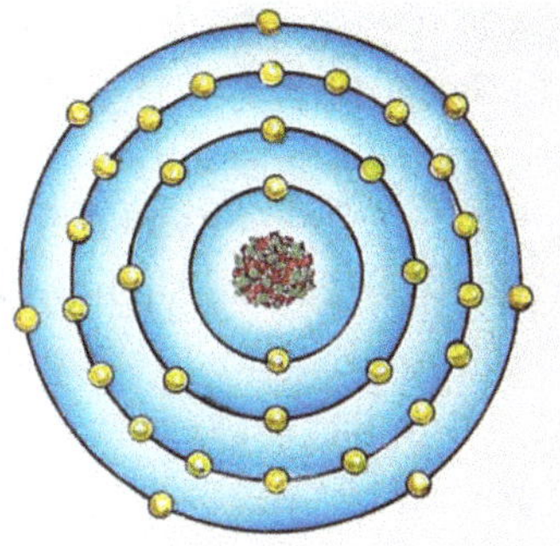

Figure B2  Electrons in shells N

Figure B2 indicates 2 electrons in shell N = 1, 8 in shell 2, 18 in shell 3 by the above formula. The 4th shell is incomplete with 7 electrons giving a total of 35 in the element bromine (Br).

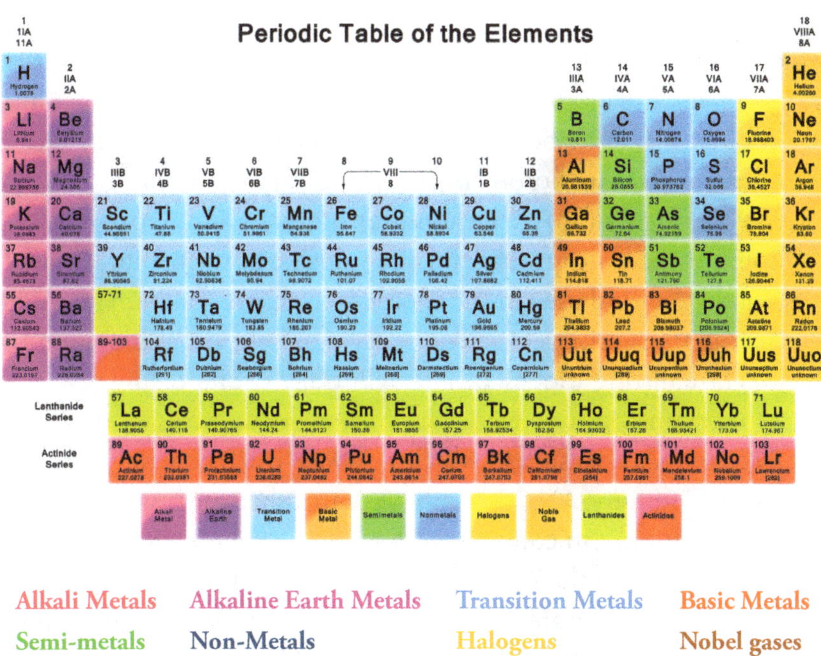



Figure B3  Periodic Table

The rows in the Table of Figure B3 are commonly called *periods* and columns are called *groups*. The elements in Group 1A, the first vertical column, each have **one** outer 'shell' electron; those of Group 2A each have **two** outer shell electrons; and on up to Group 8A which have complete outer shells of **eight** electrons each (except helium).

The abbreviations in the Table are symbols: in many cases the first letter in the English name is used H (Hydrogen), O (Oxygen), N (Nitrogen), C (Carbon), S (sulfur) and F (Fluorine). Frequently a second letter must be used as the names of more than one element have the same initial letter. The first letter is capitalized, the second is not – examples are Ca (Calcium), Co (Cobalt), Cr (Chromium), Cl (Chlorine), Cd (Cadmium). Some symbols are based upon the Latin name: Fe (Ferrum) for Iron, Cu (Cuprum) for Copper, Ag (Argentum) for Silver, Au (Aurum) for Gold, Hg (Hydrargyrum) for Mercury, Na (Natrium) for Sodium, and K (Kalium) for Potassium.

*Most of the elements are metals.* They are located on the left and center and toward the bottom of the periodic table. The nonmetals (colored dark blue) lie to the right and above the bright green boxes in the Table. The bright green boxes are the semi-metals (metal-like or metalloids) that lie between the true metals and the nonmetals – these semi-metals are boron (B), silicon (Si), germanium (Ge), arsenic (As), antimony (Sb), tellurium (Te) and polonium (Po).

Some of these are quite useful and valuable in the manufacture of semiconductors and used in the computer industry. Gallium arsenide (GaAs) can transform electricity directly into light and is used to produce light-emitting diodes LEDs. Tellurium is one of the few elements that can combine with gold. Just as tellurium is in the same family as oxygen and sulfur (which lie above it in the family of nonmetals of Column 6A) can combine with other metals to produce sulfides, tellurium can combine with gold to produce a telluride.

Column 1A are the Alkali Metals (except hydrogen which is not a metal); Column 2A are Alkaline Earth Metals; the bottom portion of Columns 3A to 6A are the Basic Metals; Column 7A are the Halogens ('salt formers') -- elements with seven outer shell electrons and are oxidizing agents capable of gaining one electron. Column 8A are the inert gases (Nobel Gases) with a full set of eight electrons in the outer shell (except for helium).

Elements tend to combine so that there are 8 electrons in the outer 'shell'. Note that sodium **(Na)** on the left has only 1 electron. If it donates 1 electron (becoming a *positive* ion) to the chlorine element **(Cl)** on the right with 7 electrons (now 8) – gaining 1 electron it becomes a *negative* ion.

The combined molecule has Na + - Cl = NaCl sodium chloride (***common table salt*** – used as a food preservative in ancient times and as an article of commerce for 1000's of years. Roman soldiers were paid with it—hence "salary" literally *"salt money"*. *Useful people were "salt of the earth" -- not so useful people were consider "not worth their salt".*

There are 118 known elements – 98 occur naturally in nature, elements 99 to 118 have been synthesized in laboratories or nuclear reactors.

As one traverses along the rows within the Periodic Table the atomic number (number of protons) and the number of electrons goes up by one unit. The protons have a charge of +1 each. The electrons have a charge of – 1 each. The element is electrically neutral with a net zero charge.

An important detail in the Table is the atomic *number* (the number in the upper left corner of an element box in the figure above. This is the number of protons in the nucleus of the element. This was 11 for sodium (Na). Usually the number of neutrons is the same as the number of protons and the atomic *weight* is the sum of the two and located at the bottom of the box. The atomic weight is 23 for Na. This means that there are 12 neutrons giving the total of 23. However, the atomic *weight* also averages in the weight of any *isotopes* of the element. Thus, the atomic weight of sodium is actually listed as 22.9. An isotope has the same number of protons and electrons as the normal element – only the number of neutrons is different. The isotope of sodium is sodium-24.

An element in the *neutral* or *elementary* state has *no overall charge*. There are as many positive protons as there are negative electrons. Carbon (with 6 protons, 6 neutrons and 6 electrons) is a special molecule. Almost all molecules in plants and animals contain carbon. It also exists in several different natural forms (called allotropes) – coal, slippery graphite and diamond (one of the hardest substances known). Carbon has an isotope (C-14 with two extra neutrons) which will be encountered in several Chapters of this book).

The element *beryllium* (atomic number 9) also has an important isotope *Be-10* which has an extra neutron. This will also appear in several Chapters as both *C-14* and *Be-10* are created by *cosmic rays* from space and their measured values imply the effectiveness of the strength of the Sun's magnetic field in keeping these *cosmic rays* at bay -- away from the Earth. The details of this process and just how these two isotopes are created by the *cosmic rays* will be revealed later.

Oxygen (O) is also a very special atom, as well as the special molecule $O_2$. Both are a part of the two gases we are concerned with in the study of climate change: water vapor $H_2O$ and carbon dioxide $CO_2$. *Both the water molecule and the carbon dioxide molecules represents an example of electron sharing.*

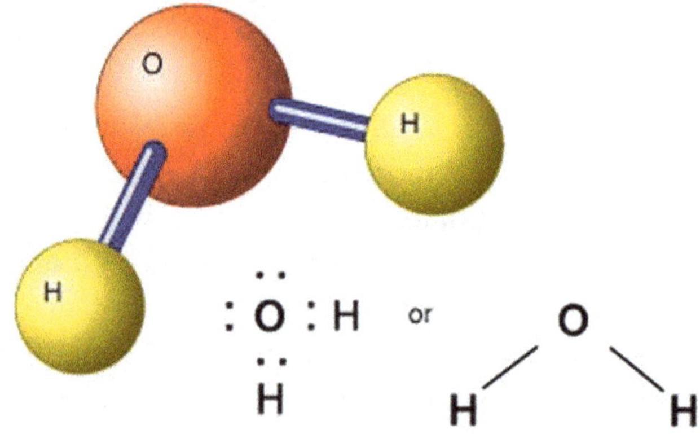

© 2006 Encyclopædia Britannica, Inc.

Figure B4  The water molecule $H_2O$

The element oxygen with *atomic number* 8 has 2 electrons in its inner shell and 6 in its outer shell – and thus shares one electron with each of 2 hydrogen atoms giving "8" in the outer shell. Its *atomic weight* is 16 grams. A number of grams of an *element* or *compound* is equal numerically to the *atomic weight* or *molecular weight* -- defined as a *mole* of that substance. *Thus, a mole of the oxygen atom is 16 grams; a mole of the molecule $O_2$ is 32 grams; a mole of a carbon atom is 12 grams; a mole of $CO_2$ is 44 grams -- considerably higher than a mole of dry air at 28.97 grams. Thus, $CO_2$ is definitely heavier that dry air*

## Carbon dioxide molecule

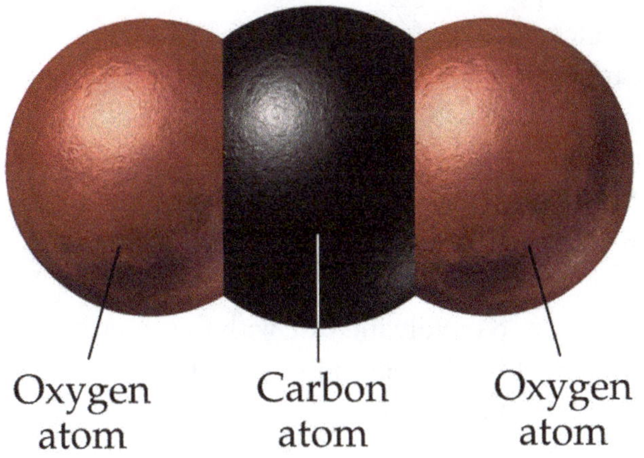


Figure B5  The carbon dioxide molecule

The carbon atom has 2 electron in the inner shell and 4 electrons in the outer shell. Oxygen atoms share 2 electrons each for 4 added to carbon's 4 giving a total of 8 in carbon's outer shell – and each oxygen atom now shares 8 electrons.

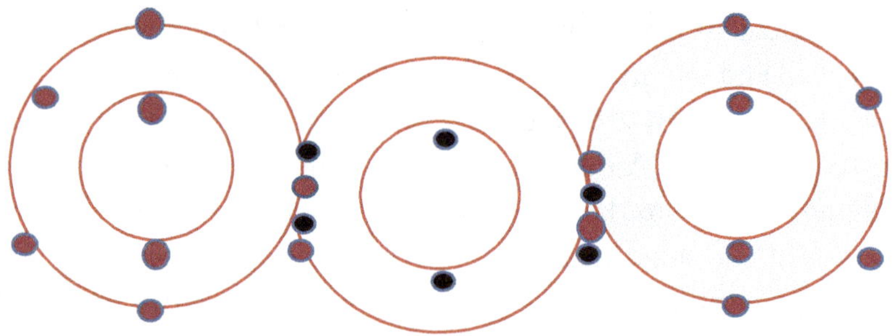

Figure B6  The carbon dioxide molecule illustrated

Oxygen is so important, one needs to say more about its evolution in the atmosphere, its significance in today's world, and of course, its role in understanding *climate-change.*

The Earth's core was 63 percent complete as an iron core and liquid iron in its first two billion years of existence. During the past 2.6 billion years of existence, the final 37 percent of the iron core was completed by the gradual gravitational differential of metallic iron from the mantle to the core. Prior to 600 Mya almost all oxygen was used for oxidizing iron.

Oxygen began to rapidly increase with the absence of iron. It also increased with multi-cell algae and increased plants on land (approximately 400 Mya.). $O_2$ then gradually increased to its present 21 percent of the Earth's atmosphere.

*$O_2$ makes up approximately one-half of our immediate environment* – it provides life: we can live weeks without food, days without water, but only a few minutes without $O_2$. The human body is composed of 65% $O_2$. The body structure of plants (cellulose, is ~ 50% $O_2$. About 46% of the rocks and soil contain $O_2$.

The Earth (crust and interior) is 32.1% iron, 30.1 % oxygen, 15.2 % silicon, 13.9% magnesium, plus the balance with other trace elements.

$O_2$ represents 21% of the atmosphere. $O_2$ is 89% of the weight of water – where the oceans cover 70% of the Earth's surface.

Another lesson about chemistry notation is given by the following simple relation:

$$\mathbf{C + O_2 \rightarrow CO_2}$$

This is not an equation; the (+) sign is read as "and" and the arrow is read "react to form" or "yield". The terms on the left side of the arrow are the "reactants" and the terms on the right side are the "products". *Equal numbers of each atom must be on both sides of the arrow.* They are the same atoms -- thus the total weight of both sides is also the same. This is the law of stoichiometry.

The above relation is an expression of the burning of coal (carbon) which is an exoenergetic reaction which means that energy is given off (the reactants have more energy than the products). However, in this case the coal must be ignited first as the bonds in the $O_2$ molecule must be broken before the bonds in the $CO_2$ molecule can be formed.

Another good example of a chemical principle is the complete oxidation of methane $CH_4$ to carbon dioxide and water.

$$CH_4 + 2\,O_2 \rightarrow CO_2 + 2\,H_2O$$

The first lesson here is that the left side of the arrow contains 1 Carbon, 4 Hydrogen and 4 Oxygen atoms, thus the right side must have the same, so there are 2 water molecules – this raises the hydrogen atoms to 4, the proper number to match the left side, but the two extra oxygen atoms required another oxygen molecule on the left as indicted.

The second lesson here is that the oxidation of methane proceeds at measurable rates at temperatures between 400 and 500 °C. However, the reaction does not take place in a single step as written above. When this reaction is observed in a closed vessel there are intermediate products that can be observed. There are more than a dozen different chemical reactions that need to be accounted for.

### *There is more to say about oxygen in the course of evaluating climate change.*

Oxygen is the third most abundant element in the universe, behind hydrogen and helium, it is one of the most important and abundant elements on Earth.

Molecular oxygen $O_2$ is produced from water by cyanobacteria, algae, and plants during photosynthesis and is part of cellular respiration for all living organisms. Green algae and cyanobacteria in marine environments produce ~70% of the free oxygen produced on Earth and the rest is produced by terrestrial plants.

The atomic number of the atom of oxygen is 8 (with 8 protons) and its atomic weight is 16 (with 8 neutrons). The conventional form of expressing atomic oxygen is $^{16}O$ – which is known as "light" oxygen. There are also a small fraction of oxygen atoms that have 2 extra neutrons and this is referred to as $^{18}O$ because of 2 extra neutrons making the atomic weight equal to 18 -- these are referred to as "heavy" oxygen. The heavy oxygen is fairly rare – found in only about 1 in 500 atoms of oxygen.

These two isotopes of oxygen are extremely important in climate analysis because of the following fact. Light oxygen in water ($H_2{}^{16}O$) *evaporates* easier than water with heavy oxygen ($H_2{}^{18}O$) – (it is harder for heavier molecules to overcome barriers to evaporation). Whereas water vapor molecules that condense and form precipitation preferentially remove $^{18}O$ relative to $^{16}O$.

When the above is applied to ice cores, the following scenario develops. The water-ice in glaciers originally came from vapor over the oceans, later falling as snow and becoming compacted in the ice. Hence glaciers are relatively enhanced in $^{16}O$, while the oceans are relatively enriched in $^{18}O$.

This imbalance is more severe in colder climates than for warmer. It has been demonstrated that a decrease of one part per million of $^{18}O$ in ice reflects a drop of 1.5 degree C in air temperature at the time it originally evaporated from the ocean. In ice cores from Greenland and Antarctica are layered and the layers can be counted to determine age – with the heavy oxygen ratio used as a thermometer.

This isotopic analysis of oxygen can be used in ocean sediment cores of the shells of dead marine organisms. The oxygen in the carbonate of the calcium carbonate ($CaCO_3$) of the organisms reflects the isotopic abundance in the shallow waters where the various sea creatures lived. Once one knows the date and time of ancient sediments, one can use the isotopic ratio of oxygen to determine sea surface temperature at that time -- the procedure of Veizer[3] who reconstructed Earth's temperature record over the past 500 million years – as discussed in Chapter 12.

Temperature is very important in this text and is exclusively used in terms of centigrade in this book (occasionally Kelvin is used --- which is just centigrade plus 273.16). The official notation for temperature is:

|  | Fahrenheit | Centigrade | Kelvin (Absolute) |
|---|---|---|---|
| Boiling point of water | 212° | 100° | 373.16° |
| Freezing point of water | 32° | 0° | 273.16° |
| Absolute Zero | - 459.7° | - 273.16° | |

# Appendix C (Quantifying the Three Forces)

The following information is repeated from Chapter 10 so that the derivation of the form of the variable $C_w$ and $C_r$ can be determined. The reference here is repeated:

Sorokhtin, O. G., G. V. Chilingar and L. F. Khilyuk, 2007: *Global warming and global cooling: evolution of climate on Earth*. Elsevier, Amsterdam, 313 pp.

$$(- \partial T / \partial z) = g / (C_p + C_w + C_r) = 6.5 \text{ °K per kilometer}$$
(equivalent to the standard lapse rate).

The required input is:

$T_E = 255$ K     the effective radiation temperature of the Earth from the Stefan-Boltzmann law

$T_S = 288$ K     the average surface temperature of the Earth

$C_p = 0.2394$     the standard value of specific heat in units of [cal / g °K]

$R = 1.987$     [cal / g °K]

$\mu = 29$ moles for the dry atmosphere

Given equation **(1)** below and the value of $\alpha$ = 0.1905 found from experimental data, the corrective coefficients $C_w$ and $C_r$ can be found.

$$(C_p + C_w + C_r) = R / \mu\, \alpha = (1.987) / (29)\,(0.1905) = 0.3597 \quad (1)$$

The equations below have been derived by Sorokhtin, et al. (see reference in text). Intermediate steps and highlighting have been added by this author for clarity.

Defining the total effective heat resource as $Q_A$ and the total effective mass of the atmosphere as $m_A$, one can represent the *radiation component* of specific heat $\mathbf{C}_r$ as a function of the effective radiation temperature $T_E$ as follows:

$$C_r = Q_A / m_A\, T_E \quad \{\text{note that these are the same units as for } C_p \text{ [cal / g }^\circ\text{K]}\} \quad (2)$$

In a similar manner one can logically consider the *additional heating* required to raise the temperature of the atmosphere from the radiation temperature $\mathbf{T}_E$ to the average surface temperature $\mathbf{T}_S$ to be determined by the corrective specific heat $C_w$ for considering water vapor condensation:

$$C_p + C_w = Q_A / m_A\, (T_S - T_E) \quad (3)$$

Using (2) in (3) one obtains $C_p + C_w = C_r\, T_E / (T_S - T_E)$, then defining $C_r$:

$$C_r = (C_p + C_w)\,(T_S - T_E) / T_E \quad (4)$$

From (1) one obtains another form for $C_r$:

$$C_r = (R / \mu\, \alpha) - (C_p + C_w) \quad (5)$$

Equating the right hand sides of (4) and (5) since they both equal $C_r$ one has:

$$(C_p + C_w) \, (T_S - T_E) \, / \, T_E = (R \, / \, \mu \, \alpha) - (C_p + C_w) \qquad (6)$$

Now in (6), gathering the $C_w$ terms on the left side and the $C_p$ terms on the right side leads to:

$$C_w \, [1 + (T_S - T_E) \, / \, T_E] = (R \, / \, \mu \, \alpha) - C_p \, [1 + (T_S - T_E) \, / \, T_E]$$

upon simplifying this becomes:

$$C_w \, [T_E + (T_S - T_E) \, / \, T_E] = (R \, / \, \mu \, \alpha) - C_p \, [T_E + (T_S - T_E) \, / \, T_E]$$

which reduces to

$$C_w \, [T_S \, / \, T_E] = (R \, / \, \mu \, \alpha) - C_p \, [(T_S \, / \, T_E] \text{ multiplying all terms by}$$
$T_E \, / \, T_S$ gives:

$$C_w = (R \, / \, \mu \, \alpha) \, (T_E \, / \, T_S) - C_p \text{ which is the final form for } C_w \qquad (7)$$

Now using (7) in (4) gives:

$$C_r = [C_p + \{(R \, / \, \mu \, \alpha) \, (T_E \, / \, T_S) - C_p\}] \, [(T_S - T_E) \, / \, T_E]$$

$$C_r = (R \, / \, \mu \, \alpha) \, (T_E \, / \, T_S) \, (T_S - T_E) \, / \, T_E]$$

$$C_r = (R \, / \, \mu \, \alpha) \, (T_S - T_E) \, / \, T_S] \text{ which is the final form for } C_r \qquad (8)$$

One can now confirm (1) above:

$$(C_p + C_w + C_r) = R \, / \, \mu \, \alpha$$

$$\{C_p + [(R \, / \, \mu \, \alpha) \, (255 \, / \, 288) - C_p] + [(R \, / \, \mu \, \alpha) \, (33 \, / \, 288)]\} = R \, / \, \mu \, \alpha$$

$$R \, / \, \mu \, \alpha = R \, / \, \mu \, \alpha$$

Therefore

$$C_p = 0.2394 \quad C_W = 0.0789 \quad C_r = 0.0414 \quad \text{-- In percent of } R \, / \, \mu \, \alpha:$$

$$C_p = 0.2394 / .3597 = 66.56\ \% \quad C_W = 0.0789 / .3597 = 21.93\ \%$$

$$C_r = 0.0414 / \quad .3597 = 11.51\ \%$$

---

The authors also have a value of $T_E = 263.6$ due to the Earth's present data precession angle.

This implies that the thermal blanket receives $299.2 - 263.6 = 24.6\ °C$ heating, not $33.2\ °C$.

Using this value rather than $T_E = 255\ K$ changes the $C_W$ and $C_r$ values as follows:

$$C_p = 0.2394 / .3597 = 66.56\ \% \quad C_W = 0.0896 / .3597 = 24.91\ \%$$

$$C_r = 0.0307 / .3597 = 8.53\ \%$$

There is another interesting calculation that is provided in the reference that uniquely determines the degree of temperature increase with a *doubling of CO$_2$*. Several authors have made an estimate of this effect – without the calculations shown in Chapter 11 of this text. This is unique as it includes the effect of more mass in the atmosphere due to more $CO_2$.

One begins by using their equation number (3.9) in the reference: $T = 288.2\ (P/ P_0)\alpha$       (9)

Where $P$ = pressure, $P_0$ = surface pressure, and $\alpha = 0.1905$ from above.

Taking the log of both sides of (9) one has:
$\ln T = \ln 288.2 + \alpha \ln P$       (10)

Introducing the notation $A(\alpha, P)$ for the right side of (10) one obtains sensitivity functions of the ln T as partial derivative of $A(\alpha, P)$ with respect to the parameters $\alpha$ and P.

Differentiating (10) one obtains:
$$(1 / T) \, dT = [\partial A(\alpha, P) / \partial \alpha] \, d\alpha + [\partial A(\alpha, P) / \partial P] \, dP \qquad (11)$$

Substituting the partial derivative of $A(\alpha, P)$ and multiplying both sides of (11) by T gives:

$$dT = T \ln P \, d\alpha + (\alpha / P) \, dP \qquad (12)$$

Replacing the differentials by *finite differences* one has approximately:

$$\Delta T = T \ln P \, \Delta\alpha + T (\alpha / P) \, \Delta P \qquad (13)$$

Doubling $CO_2$ changes $\Delta P$ by $1.48 \times 10^{-4}$ atm and $\Delta\alpha = -4 \times 10^{-6}$. At sea level if pressure is measured in atmospheres, then P = 1 and ln P = 0, therefore (13) becomes:

$$\Delta T = T \, \alpha \, \Delta P = (288.2) \, (0.1905) \, (1.48 \times 10^{-4})$$
$$= 8.12 \times 10^{-3} = < 0.01°C$$

If one goes higher in the atmosphere, say h = 10 km, then the pressure and temperature drop and the numbers become ~ P = 0.24 atm, ln P = -1.4286, $\Delta P$, $\alpha$, and $\Delta\alpha$ do not change and

$$\Delta T \sim = 2.710 \times 10^{-2} = < 0.03°C$$

*Both of these numbers are well below what the IPCC declares for the doubling of $CO_2$.*

These authors also point out that after the year 2100, of that anthropogenic increase of $CO_2$ the portion which dissolves into the ocean would lead to less carbon and oxygen in the atmosphere and a slight reduction in atmospheric pressure -- producing a slight cooling by perhaps ~ - 0.01°C. "In reality, however, the metabolism of plants should almost completely compensate for the disruption of equilibrium of mankind and restore the climatic balance."

# Appendix D (Radiation Details)

The Electromagnetic Spectrum with the seven major wave regions from radio to gamma-rays. Radiation is energy traveling at the speed of light in the form of particles *(photons) without mass* traveling as a stream in a wave-like pattern. The waves can be measured in terms of wavelength (which for our discussion for $CO_2$ and $H_2O$ is in microns or $10^{-6}$ meters), their frequency (cycles per second or Hertz) or by the amount of energy in the photons measured in terms of electron volts (one electron-volt = $1.6 \times 10^{-19}$ Joules). The photon contains the energy.

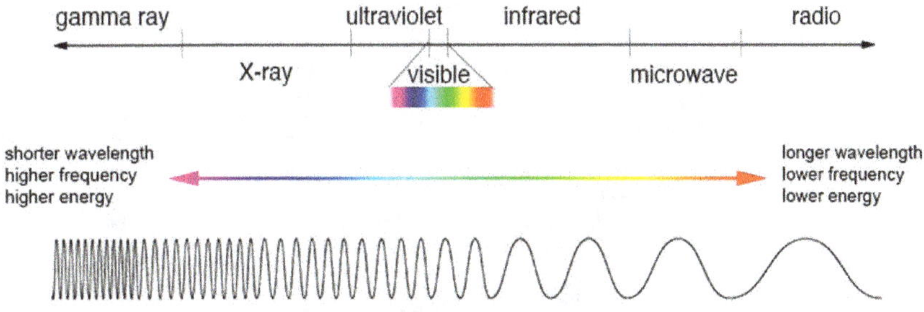


Figure D1  The Electromagnetic Spectrum

As indicated in the figure (produced by NASA) the *longest radio* waves on the far right have the lower frequency and lower energy. The *shortest gamma-rays* on the far left have the highest frequency and the highest energy.

Each group of waves are not fundamentally different, but each is produced by a different process and each is detected by a differet method. The electromagnetic waves move at the speed of light which by manipulating the Maxwell equations, it can be shown that this is given by:

$$\mu_0 \varepsilon_0 = 1.112 \times 10^{-17} \left[ s^2 \middle/ m^2 \right];$$

$$C = \text{Speed of light} = \frac{1}{\sqrt{\mu_0 \varepsilon_0}} = 3 \times 10^8 \ m/s$$

The electromagnetic field is composed of the *electric field (in red)* and the *magnetic field (in blue)*. The classical pespective is that these fields are thought to be produced by smooth motions of charged particles which are oscillating charges that produce electric and magnetic fields that may be viewed in a contiuous fashion. The wave below is propogating in the Y-direction.

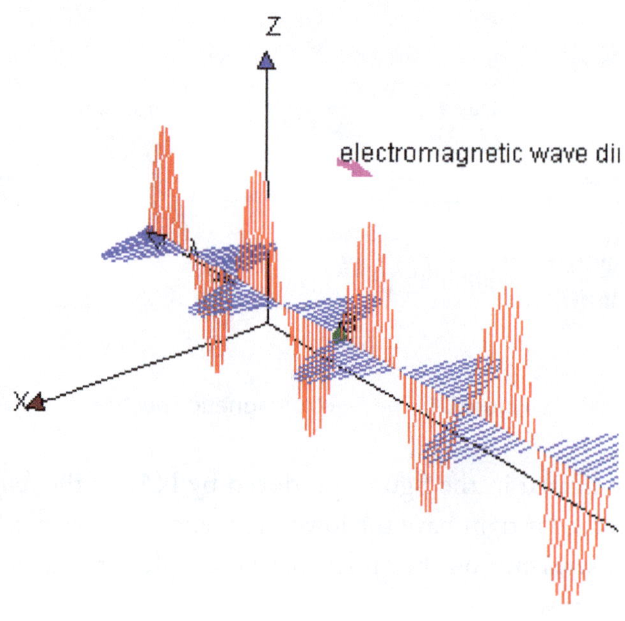

The Sun produces a continuous stream of solar radiation on the rotating Earth each day. The solar energy absorbed by the Earth's surface is returned to space by longwave thermal radiation emitted from the surface of the Earth and from the atmosphere.

Very little of the original surface heat intensity actually reaches the stratosphere above the tropopause, the top of the troposphere. The stratosphere is also in thermodynamic equilibrium. The source of the heat in the stratosphere is the *direct absorption of solar energy by ozone.* The cooling of the stratosphere for balance is due to longwave emission primarily by $CO_2$ and to a lesser extent by water vapor and ozone. The reader can consult Houghton or Liou for details on radiation. Their references are listed in Chapter 11.

Radiation transitions between molecules must couple with an electrodynamic field so that energy exchanges can take place. If the effective centers of the positive and negative charges of the molecule have a separation which is nonzero, then an electric dipole exists for energy transitions. The radiative gases in the infrared of $H_2O$ and $CO_2$ have permanent electric dipole moments due to their asymmetric charge distributions. These gases participate in the radiation process – they absorb and emit energy – *they do not create energy* – they simply move it about.

The atmospheric gases nitrogen ($N_2$) and oxygen ($O_2$) have linear molecules thus they have a *symmetric charge distribution and are inactive* in the *infrared* region of the electromagnetic spectrum. However these molecules do have weak *magnetic* dipole moments that allow radiative activities in the ultraviolet region of the spectrum.

Here the most important equation is Planck's function for radiation intensity:

$$B_\lambda = [2\,h\,c^2\,/\,\lambda^5]\,[\exp\,(ch\,/k\,\lambda\,T) - 1]^{-1} \tag{1}$$

The form of the Planck intensity of radiation used in Chapter 11 is that of (1). It is a strong function of both temperature and wavelength – as indicated in a Fig 11.2. $B_\lambda$ expresses the power (watts per m$^2$) per unit wavelength interval per unit solid angle. T is temperature. Integrating $B_\lambda$ over the entire wavelength domain from 0 to infinity gives B(T) = ~ T$^4$ . Noting that B(T) is independent of direction, it can be shown that integrating over an entire hemisphere leads to $\prod$ B(T) = (5.67 x10$^{-8}$ W m$^2$ K$^{-4}$) T$^4$ = flux density (energy per unit area per unit time).

The wavelength line shape depends upon the type of broadening – in the troposphere it is primarily pressure broadening (collision with nitrogen and oxygen) and the Lorenz line shape applies for the absorption coefficient K ($\nu$) or K ($\lambda$).

K ($\nu$) = [S / $\prod$] {$\alpha$ / [($\nu$ - $\nu_o$) $^2$ + $\alpha^2$]} where $\alpha$ = $\alpha_o$ (**P** / P$_o$) [T$_o$ / **T**] $^{1/2}$ and where *S = the line strength,* $\nu_o$ is the central frequency of the line, and $\alpha_o$ is the half-width (one half of the width of the line at the level where k ($\nu$) is one half of its value at the line center).

The line strength is defined as the integral of k d$\nu$ = [S = $\int$ k d$\nu$] and increases with decreasing T. The change in line shape with height is affected by P and T. Calculations in Chapter 11 have been performed with temperature and pressure affects due to broadening with slight changes in the numerical values, *but no change in the height where the intensity becomes trivial.*

*A pencil of radiation* traversing a medium will be weakened by its interaction with matter, If the intensity of radiation I$_\lambda$ becomes I$_\lambda$ + d I$_\lambda$ after traversing a thickness ds in the direction of propagation, then

dI$_\lambda$ = - k$_\lambda$ $\rho$ I$_\lambda$ ds or *dI$_\lambda$ / k$_\lambda$ $\rho$ ds = - I$_\lambda$ the Beer-Bouger-Lambert law* (2)

where $\rho$ is the density of the material and k$_\lambda$ is the mass absorption coefficient (*assuming diffuse radiation from multiply scattering is neglected*).

Now consider the transfer of thermal infrared radiation emitted for the Earth and the atmosphere where a beam of intensity will undergo the *absorption* and *emission* processes simultaneously. The *Schwarzschild equation* for this process is: *dI$_\lambda$ / k$_\lambda$ $\rho$ ds = - I$_\lambda$ + B$_\lambda$ (T) where the first term on the right is the reduction due to absorption; the* *second term represents the increase of* radiant intensity arising from the blackbody emission of the material.

One can derive the *Solution #1 algorithm* from Liou's definitions/ integrals for level i:

$$T_\lambda (\tau) = e^{-\tau} \, ; \, d \, T_\lambda (\tau) \, / \, d \tau = -1 \, e^{-\tau}; \, \tau^{sfc} = \tau^*;$$

$$B(sfc) = B(0) = F(I\text{-}1); \, \Delta \, \text{tau} = \Delta \tau = \rho_i K_\lambda \Delta z_i$$

$$F(1) = B(\tau^*) \, T_\lambda (\tau^* - \tau) - \int_\tau^{\tau^*} B(1) \frac{d}{d\tau'} \, T(\tau' - \tau) \, d\tau';$$

$$C1 = [e^{-\Delta \tau}] \text{ and } C2 = [1 - e^{-\Delta \tau}].$$

$$= B(0) \, e^{-\Delta \tau} - B(1) \, d/d\tau \, T_\lambda (\tau - \tau) - B(1) \, e^{-\Delta \tau}$$

$$= B(0) \, e^{-\Delta \tau} + [1 - e^{-\Delta \tau}] \, B(1)$$

$$\{\text{since } d \, T_\lambda (\tau) \, / \, d \tau = -1 \, e^{-\tau} \text{ and } e^0 = 1\}$$

$F(I) = [e^{-\Delta \tau}] \, F(I\text{-}1) + [1 - e^{-\Delta \tau}] \, B(I)$. The algorithm using this result is found in Chapter 11.

Solar radiation comes from a very distant point, the Sun, thus it can be treated as parallel unidirectional radiation – *a pencil* of radiation like stated above. However, terrestrial radiation comes from all directions since each molecule acts as an individual 'very small sun' for thermal *diffuse* radiation. Thus thermal radiation emitted by the Earth and the atmosphere comes from all directions. However, if the atmosphere is considered to be plane parallel, only changes of parameters in the *vertical* need to be considered and the diffuse nature of the beam can be considered by a diffuse transmissivity as shown below.

*A differential optical depth can be defined as $d \, \tau = - k_\lambda (z) \, \rho (z) \, dz$.*
*A Figure from Liou[1] was shown in the text of Chapter 11. It would also apply here.*

Houghton has an equation [(5.15) on page 155] for the vertical divergence of the net flux:

In the formula: B = Planck function = B $(\lambda, T)$, T = temperature, $\tau^*$ (tau) = diffuse transmissivity

$dF_{NET}/dh = -\int (dB/dT)(dT/dh)(d\tau^*/dh)\,dh + \int (dB/dT)(dT/dh)(d\tau^*/dh)\,dh$ (4)

where the limits of integration on the first integral are top to bottom:

h = 0 and $h_t$ (r for reference height)

and where the limits of integration on the $2^{nd}$ integral are top to bottom:

$h_t$ -- t for top and $h_r$.

One can compute the layer changes in $\Delta$ tau directly from Houghton's equation without the integration factors of $[e^{-\Delta\tau}]$ and $[1 - e^{-\Delta\tau}]$ and obtain just slightly different answers (< 2%).

The solution to reducing the diffuse radiation to a simple geometric form was found empirically by assuming the diffuse transmissivity $\tau^*(u)$ is related to the parallel beam transmissivity by $\tau^*(u) = \tau$ (1.66u). Both Houghton and Liou agreed on this. From the exponential form shown in the text of Chapter 11, the diffusion factor used here will be the same value of 0.811124.

One can solve these equations in a second way based upon Houghton's equation. *This is called Solution #2.* Here the absorption coefficients must be per km (they were provided to the author as per cm and were multiplied by $10^5$ to apply to changes per km.). Thus the maximum coefficient in Band 1 over the range of wavelengths of 1.0 to 4.50346463 is $K_\lambda = 4596$ m$^2$ / kg.

The calculations were performed both with and without the influence of line shape. Line shape is only important only for *strong lines* and was only used *when $K_E$ was > 50*. Adding the line shape changed the numbers slightly, but not the end result – line shape calculations are excluded here.

Calculations begin with the $CO_2$ lines and coefficients in Band 1. The maximum coefficient in this band has the large value $K_\lambda$ = *4596 $m^2/kg$* and occurs at the wavelength of $\lambda$ = *4.2346463 $\mu m$*. All 390,000 lines are organized by increasing wavelength, paired with their surface absorbance coefficient. One starts with a formula for the lines. The number of lines *displayed* in this Appendix vary from 25,001 to 70,001. Lines up to 300,000 were run in the author's previous paper published in March of 2018 – see reference in Chapter 11. *Many more runs were performed over a wide range of lines – all of these runs provided fairly similar numbers with the same basic conclusion.*

Band 1 run uses the line formula with J = 1 to 70001: $\lambda$ = *1.0 + (J – 1) * 0.00005 + 0.0346463*. This provides lines from 1.0 to 4.50346463 $\mu m$. The value of J = 64001 provides a direct hit on the largest absorption coefficient at $\lambda$ = 4.2346463.

Every line in the formula is evaluated at every level ($\Delta$ height = one km) with the following steps for each line and level. Information for each level is saved going upward, then downward.

(1) The line is selected from the formula for J = 1 to 70,001
Note that the only line guaranteed to be *exact* is $\lambda$ = 4.2346463. Other lines *may* also be exact, but in any case they are so extremely close that linear interpolation provides the proper coefficient.

(2) The standard temperature T at that height is selected for use in the Planck subroutine along with the wavelength to arrive at the proper Plank radiation intensity for that B $(\lambda, T)$.

(3) Rather than separate each reduction with height -- i.e., the reduction due to the Planck function B $(\lambda, T)$, the reduction due to changes in the $CO_2$ density; and the reduction to the

coefficient magnitude with temperature, one can obtain the same final answer by *accumulating* those changes into a *single coefficient of reduced intensity ($K_E$)*. The first of the three steps is the Planck change with height: $K_E = K_\lambda \times [B\,(\lambda,\,T)\,/\,B\,(\lambda,\,T_{surface})]$, the original surface $K_\lambda$ is used for the first level.

(4) The new $K_E$ is further reduced by the density change in $CO_2$. [Other versions in Chapter 11]
$K_E = K_E \times \rho\,(T,\,P)_{surface}\,/\,\rho\,(T,\,P)$ where $\rho = P\,/\,1.889\,T$. [Where $P = \rho\,R\,T$, and R is for $CO_2$.]

(5) The new $K_E$ is further reduced by the decreasing temperature with height as $K_\lambda$ increases in line strength with height. $K_E = K_E \times (T\,/\,T_{SURFACE})$.

(6) The final step is to correct for the path length since the radiation is diffuse. Computing the upward flux to at given layer using the Level 1 as the base rather than the surface, provided an estimate of the downward flux to be expected. The results of *Solutions 1 and 2* are very similar.

Having performed all the steps above, the data is saved, checked and statistics determined for each level: including the percent transparent determined for all $K_E < 1$. Note that K in all the Tables is

$K_\lambda$ at the surface and $K_E$ at all layers above the surface.

The results for Band 1 (Table 1) reveal several important points. The first is that 93.68% of the surface coefficients $K_\lambda$ are transparent with values < 1 – quite a large number. This is important as *all the $CO_2$ molecules are influenced by all the coefficients.* [Note that 98 % of the surface coefficients were transparent between 1 and 40 µm as indicated in Table 11.2]

The derived coefficients of reduced intensity are 100% transparent from *12 km and upward.* Despite the very powerful absorption coefficients in Band 1 (there were 102 values greater than 1000), *the influence of the Planck function is very strong – especially at the shorter wavelengths of Band 1 (see Table 11.3 in text.*

*Solution #1* has a problem with *very large absorption coefficients properly.* For *Solution #1* the absorption coefficients must be divided by 1000. Compare the results in Table D1 where the absorption coefficient has the value 4.596 for Solution #1 and the value 4596 for Solution #2.

The maximum absorption coefficient in Band 1 is 4,596 as required for *Solution #2.* For **Solution #1** the maximum coefficient is reduced by 1000 and has the value 4.596. Both Solutions have small Net Planck Intensity values as they should at this wavelength, but the values for *Solution #1* are extremely small – already at 3km as shown in Table D1.

The problem arises as the absorption coefficient of 4.596 is so strong that the exponential parameters C1 and C2 are at their limits over much of the atmosphere. C1 has the value of ~ 0 from the surface to 10 km and only slowly *increases* to the value of 0.185 at 18 km. Thus C2 has the value ~ 1.0 from the surface to 10 km and then slowly *decreases* to 0.815. Thus, *Solution #1* returns a Net of ~ 0 at levels from 1 to l0 km, and the result in Table D1 is 3 km for Solution #1.

This is not a credible result – the exponential terms mask the detail. However, the use of *Solution #2* here provides the value of $K_E$ reaching a *critical value of < 0.333 – at 11 km. This provides a clear level where the $CO_2$ radiation intensity is virtually transparent.* This value of $K_E$ provides a clear metric that the proper level is 11 km. It also provides the Planck intensity at that level of **0.83 x 10$^{-3}$** -- which comes very close to the risen value of **0.165 x 10$^{-3}$** from the *Solution #1.*

Thus, our perspective is that the *Solution #2,* which gives answers within 2 % of *Solution #1* for all the coefficients – but *Solution #2* is the slightly better solution which provides more meaningful information for all values of the absorption coefficients.

Table D1  Hybrid results for Band 1 using both Solution #1
and Solution #2

| Line | 70001 | 70001 | 70001 | 70001 | 70001 | 70001 |
|---|---|---|---|---|---|---|
| K = coefficient | Surface | 3 km | 5 km | 7 km | 9 km | 11 km |
| % Transparent | 93.68 | 99.31 | 99.66 | 99.86 | 100 | 100 |
| Max K coefficient | 4596 | 41.5 | 15.5 | 4.28 | 0.932 | 0.161 |
| Avg. K coefficient | 5.722 | 0.052 | 0.019 | 0.0054 | 0.0012 | 0.0002 |
| Max Planck Net intensity | 0.6680 | $.218 \times 10^{-3}$ | $.150 \times 10^{-3}$ | $.994 \times 10^{-4}$ | $.632 \times 10^{-4}$ | $.381 \times 10^{-4}$ |
| Max Planck Net intensity | 0.6680 | 0.073 | 0.030 | 0.011 | 0.0032 | $0.86 \times 10^{-3}$ |
| K < .0001 | 26410 | 58505 | 61275 | 63913 | 65457 | 66646 |
| .0001 ≥ K < .001 | 22089 | 4391 | 3486 | 2062 | 1539 | 2082 |
| .001 ≥ K < .01 | 9236 | 2922 | 1859 | 1653 | 2142 | 1027 |
| .01 ≥ K < 0.1 | 4567 | 1822 | 2170 | 1877 | 689 | 218 |
| 0.1 ≥ K < 1.0 | 3273 | 1883 | 975 | 397 | 174 | 28 |
| 1.0 ≥ K < 10 | 1915 | 384 | 214 | 99 | 0 | 0 |
| 10 ≥ K < 100 | 2002 | 94 | 22 | 0 | 0 | 0 |
| 100 ≥ K< 1000 | 407 | 0 | 0 | 0 | 0 | 0 |
| K ≥ 1000 | 102 | 0 | 0 | 0 | 0 | 0 |

Table D1 has the entire *Solution 2,* with the spectrum of K values according to size are indicated in each layer. Just one line of the Net Planck Intensity is shown for *Solution #1.*

When the absorption coefficients are extremely small, *Solution #1* does also provide Net radiation which is ~ 0. This was shown in the

Text of Chapter 11 with a zero coefficient. This algorithm works well at all coefficients except the very large. It may also work well if the vertical resolution of the integration is over a finer vertical resolution than the one kilometer resolution used in this study.

Runs for Band 2 are shown for *Solution #2* with the maximum coefficient 596.1 in Table D2. The results in Table D2 indicates two runs of 25,001 and 50,001 lines over the wavelength region indicated.

Table D2.  Solution #2 over Band 2 with 25,001 and 50,001 lines over 7.98133 to 17.98133 μm

| Lines Schwarzschild | 25,001 | 25,001 | 25,001 | 50,001 | 50,001 | 50,001 |
|---|---|---|---|---|---|---|
| K = Coefficient | Surface | 8 km | 16 km | Surface | 8 km | 16 km |
| % Transparent | 87.940 | 99.824 | 100 | 87.938 | 99.820 | 100 |
| Max K | 596.1 | 2.67 | 0.326 | 596.1 | 2.67 | 0.326 |
| Average K | 2.206 | 0.0098 | 0.0012 | 2.206 | 0.0098 | 0.0012 |
| K < .0001 | 2125 | 16308 | 19394 | 4258 | 32610 | 38776 |
| .0001≥ K < .001 | 8931 | 3367 | 3511 | 17846 | 6732 | 7028 |
| .001 ≥ K <.01 | 3974 | 3469 | 1675 | 7962 | 6926 | 3344 |
| .01 ≥ K < .1 | 3691 | 1476 | 367 | 7361 | 2971 | 744 |
| .1 ≥ K < 1. | 3265 | 337 | 54 | 6543 | 672 | 109 |
| 1. ≥ K < 10. | 2395 | 44 | 0 | 4791 | 90 | 0 |
| 10 ≥ K < 100. | 493 | 0 | 0 | 989 | 0 | 0 |
| 100 ≥ K < 1000 | 127 | 0 | 0 | 251 | 0 | 0 |
| K > 1000 | 0 | 0 | 0 | 0 | 0 | 0 |
| Total Lines | 25,001 | 25,001 | 25,001 | 50,001 | 50,001 | 50,001 |

Above the surface, the K's in these Tables are the *single coefficient of reduced intensity ($K_E$)*.

Note that in both runs greater than 87.9 % of the surface absorption coefficients are < 1.0

Both of the runs in Table D2 provide the same answers. Many slightly different runs with different temperature profiles provided a critical value for the coefficient of reduced intensity (**$K_E$**) of approximately **0.33**. It can be seen that both runs gave the same $K_E$ value of 0.326 – implying that the level of *virtual transparency* was at 16 km.

The value of the Net Planck intensity at 16 km from both runs was 0.1262 which gives the reduction ratio of 0.1262/5.8527 = 2.16 %. Thus the Planck intensity at the surface is reduced by the factor of 100/2.16 = approximately 46.3 for the largest coefficient in this region at 16 km. All the other lessor valued surface coefficients are reduced to extremely small values and are totally transparent. These final values are indicated by the spectrum of results in the various size categories. Note that the shape of the spectrum is the same in both runs at each of the levels. Though, of course, the values are approximately *double* at each level in the second run since there are *twice as many lines* in the second run as the first run.

The final Table here indicates the results for *Solution #1* compared to *Solution #2* for the important Band 2 which includes the maximum $CO_2$ absorption coefficient at 14.98133 μm with the value of 596.1 m²/kg (for *Solution #2*) and 0.5961 for *Solution #1*). The Table is complete for *Solution #2*, with all 25,001 lines accounted for. However, Table D3 shows only the one line for the Net Plank intensities at the *surface, 8 km and 16 km* for *Solution #1*.

Table D3  Complete **Solution #2** / with **Solution #1** for lines 7.98 to 17.98

| *Lines Schwarzschild* | *25,001* | *25,001* | *25,001* | *Percent level 16 km over SFC* |
|---|---|---|---|---|
| *K = Coefficient* | *Surface* | *8 km* | *16 km* | |
| *% Transparent* | *87.940* | *99.824* | *100* | |
| *Max K* | *596.1* | *2.67* | *0.326* | |
| *Net intensity at Level Solution #1* | *5.8527* | *0.3668* | *0.2069* | *3.535%* |
| *Net intensity at Level Solution #2* | *5.8527* | *0.3194* | *0.1262* | *2.156%* |
| *Average K* | *2.206* | *0.0098* | *0.0012* | |
| *K < .0001* | *2125* | *16308* | *19394* | |
| *.0001≥ K < .001* | *8931* | *3367* | *3511* | |
| *.001 ≥ K <.01* | *3974* | *3469* | *1675* | |
| *.01 ≥ K < .1* | *3691* | *1476* | *367* | |
| *.1 ≥ K < 1.* | *3265* | *337* | *54* | |
| *1. ≥ K < 10.* | *2395* | *44* | *0* | |
| *10 ≥ K < 100.* | *493* | *0* | *0* | |
| *100 ≥ K < 1000* | *127* | *0* | *0* | |
| *K > 1000* | *0* | *0* | *0* | |
| *Total Lines* | *25,001* | *25,001* | *25,001* | |

The value of *the coefficient of reduced intensity* has achieved the low value 0.326 which is less than the critical value of 0.33, deemed to indicate the level at which the Planck intensity has reached a sufficiently low point – allowing the trace amount of heat remaining to pass upward to space – the absorption coefficient has passed from 596.1 to 0.326—dropping by a factor of 1828 and the Planck intensity has dropped by a factor of 100%/ 2.156% equals a factor of > 46. The values from ***Solution #1*** are only 0.05 less at 8 km and 0.08 less at 16 km. *Both answers from the two Solutions are virtually the same. The key point of **Solution 2** is that the metric* $K_E$ *determines the upper atmospheric* level that indicates the trivial transparency.

Back radiation where the **net flux at a level** is the upward flux at the bottom of a layer **minus** the downward flux at the top of a layer can be confusing for some -- as it leads to greater intensity at the surface. Thus, a short demonstration is provided below. A back radiation demo is found at www.ssec.wisc.edu/library/coursefiles/03_abs_em. See the cartoon below of a simple atmosphere with two layers above the surface with the longwave absorption/emission having a coefficient aL = a = 0.5. $Y_S$, $Y_M$ and $Y_U$ are radiation sources from levels Surface, Middle and Upper levels of the hypothetical atmosphere.

The incoming surface radiation on any given day is E.

$\downarrow$ E $\uparrow$ $(1-a)^2 Y_S$ $\uparrow$ $(1-a) Y_M$ $\uparrow$ $Y_U$

------------------------------------------------------------- Upper atmosphere

$\downarrow$ E $\uparrow$ $(1-a) Y_S$ $\uparrow$ $Y_M$ $\downarrow$ $Y_U$

------------------------------------------------------------- Middle of atmosphere

$\downarrow$ E $\uparrow$ $Y_S$ $\downarrow$ $Y_M$ $\downarrow$ $(1-a) Y_U$

------------------------------------------------------------- Surface of Earth

One can follow the radiation paths by the arrows. On the left, the radiation from the Surface goes up at full intensity, it is received at the Middle level at 1/2 intensity (since a = 0.5), and sent further upward and received at the Upper level at $(1-a)^2$ at 1/4 intensity.

In the Middle level the radiation in the Middle is sent up and down. The radiation from the Middle is received at the Upper level at 1/2 intensity of what the Middle received.

The radiation at the Upper level is sent upward, and downward to the Middle level, and further downward to the Surface at intensity (1-a).

The radiative balance for each of the three surfaces requires the following three equations:

$$0.25\ Y_S\ +\ 0.5\,Y_M\ +\ \quad Y_U = E \quad \text{Equation 1 (Upper)}$$

$$0.5\ \ Y_S\ +\ \quad Y_M\ -\ \quad Y_U = E \quad \text{Equation 2 (Middle)}$$

$$\quad Y_S\ -\ \quad Y_M\ -\ 0.5\,Y_U = E \quad \text{Equation 3 (Surface)}$$

The solar energy/heat reaches all the layers; the solar energy absorbed by the Earth is considered negligible with $a_s = 0$. One can now compute the fraction of longwave energy (E) received by each surface. The *total energy E is not changed* by the amounts of water vapor or carbon dioxide, but the fraction of E at a given level can change due to radiative transfer.

These three equations with three unknowns are easily solved. The can be solved by several methods to produce the answers:

$$Y_S = 1.6666667\ E \quad Y_M = 0.5\ E \quad Y_U = 0.3333333\ E$$

Thus, because the radiation is diffuse in all directions, back radiation allows the surface to receive a greater percentage of E. However the energy is still conserved as that extra heat was extracted from higher levels. This occurs in the total troposphere as energy is conserved by the three processes described in Chapter 10. One can check these answers in each equation:

Eq. 1: $\{(0.25)\ (1.6667) + (0.5)\ (0.5) + (1)\ (0.3333)\}$
$E = [0.416667 + 0.25 + 0.3333]\ E = 1\ E$

Eq. 2: $\{(0.5)\ (1.6667) + (1)\ (0.5) - (1)\ (0.3333)\}$
$E = [0.8333 + 0.5 - 0.333]\ E = 1\ E$

Eq. 3:  {(1) (1.6667) - (1) (0.5) - (0.5)(0.333)}
E = [1.6667 - 0.5 - 0.1666] E = 1 E

# Appendix E (Svensmark's Success)

A complete history of the trials and tribulations of Henrik Svensmark in producing a theory of cosmic rays interacting with the Earth's atmosphere is contained in the book of Henrik Svensmark and Nigel Calder "The Chilling Stars". This Appendix provides just a few details from that reference needed to accompany our book here -- readers are strongly encouraged to read "Chilling Stars" for the complete picture.

The Earth's climate has always changed in various degrees. The last *really significant* change encountered by modern man was The Little Ice Age from 1303 to 1850, which was well documented and had a severe impact on society. The sunspot record for the period (see Fig 5.2 in the text) convinced many to seek proof that the Sun was responsible for climate change.

However, the solar irradiance varies little over a complete sunspot cycle and less was known about the magnetic field of the Sun at that time. We now know that the intensity of the magnetic field in a typical place on the surface of the Sun is around one Gauss, about twice as strong as the average field on Earth (~ 0.5 Gauss). However, since the Sun's surface is over 12,000 times larger than the surface of the Earth and the overall influence of the Sun's magnetic field is vast! Moreover, near a large sunspot on the Sun, the magnetic field can be as large as 4,000 Gauss.

While other theories evolved about climate change, Danish scientists, led by Henrik Svensmark, did think that cosmic rays could have a more direct effect on the climate by virtue of their ability to form clouds and cool the planet. The Earth's magnetic field can shield some lessor effects from the Sun. The Sun's magnetic field is much stronger, and when it is strong it can shield cosmic rays from reaching Earth. However, when the Sun's magnetic field weakens, cosmic rays do reach Earth's atmosphere and are too strong for Earth's magnetic field to shield them from the atmosphere. Ultimately, Svensmark succeeded in proving that only the most energetic charged particles due to the impact of incoming cosmic rays can reach sea level on Earth.

Cosmic ray protons (as described in Chapters 5 and 12) on the production of the cosmogenic radionuclides carbon-14 ($^{14}$C) and beryllium-10 ($^{10}$Be)) can not only disrupt molecules and atoms, but their collisions also produce muons (heavy electrons). The muon is a negatively charged particle like the electron, but is about *200 times heavier than the mass of the electron.*

The Svensmark team showed that it was the muon that helped produce more clouds and that almost all the muons reaching the lowest 2000 meters of the atmosphere are products of incoming particles *too energetic to be affected by changes in the Earth's magnetic field.* On the other hand the *Sun's magnetic field* (significantly stronger than the Earth's magnetic field) provided a shield that reduced the cosmic rays and clouds in the 20th century – providing the increased *warming since* the Little Ice Age. The clouds we speak of here are the low clouds (at altitudes < 3000 meters) that cover huge areas of the Earth, particularly over vast ocean areas. What makes the clouds form?

Prior to 1996 textbooks indicated that when humid air becomes cold enough, the moisture can condense and make clouds. However, first there had to be small specks in the air, cloud condensation nuclei (ccn) on which the water droplets can form. The most important and the most common ccn are themselves droplets, made from molecules of sulfuric acid and water. These *molecules also needed to be seeded* and how this occurred was a mystery.

A research aircraft flying at a level of 160 meters over the Pacific Ocean in 1996 discovered a new phenomenon. But before discussing the mystery, the phenomena and its solution, one must discuss the cloud condensation nuclei.

Seventy percent of the Earth's surface is covered by vast oceans. Microscopic plants drift as plankton on these waters (dinoflagellates, plankton, and so on) When grazing creatures rupture these plant cells and microbes break down their contents, dimethyl sulfide is a product. The ocean releases large amounts of sulfur into the lower atmosphere via this dimethyl sulfide vapor – a blend of two carbon and six hydrogen atoms and one sulfur atom. Birds follow the scent of dimethyl sulfide (with the smell of shoreline seaweed) over the ocean to find food; as the day wears on the scent fades as chemical action in the air, driven by solar rays, converts the dimethyl sulfide into minute droplets of sulfuric acid.

Other contenders for the role of ccn are dust and grains of pollen, but they are generally too coarse to be efficient ccn. Over ocean areas, sea salt is the chief rival of sulfur as a provider of cloud condensation nuclei. Grains of sodium chloride of a suitable size are thrown into the air from breaking storm waves from high winds from the Roaring Forties – primarily in the winter. But these are estimated to only contribute about 10% of the required cloud condensation nuclei.

Returning to the fact that the sulfuric acid droplets are the most common ccn, what is the source of the sulfur? Over land there is a 100 million tons of sulfur produced from developed and developing countries in their various manufacturing processes. From this and other sources, the industrialized regions spread these particles downstream. Over vast areas of the open oceans, covering more than half of our planet, cloud-making relies on the sulfuric acid droplets made from dimethyl sulfide. Although this ocean sulfur is less than half of that produced over land, the ocean sulfur participates in the weather over a much greater area.

Now returning to the puzzle: when cloud condensation nuclei that seed the formation of water droplets, are themselves droplets of other vapors such as sulfuric acid, how do they form?

The conventional theory before 1996 relied on high concentrations of sulfuric acid molecules in vapor form. These should recruit a few necessary water molecules and then *slowly club together in droplets*. This theory died from the aircraft results, flying above the Pacific Ocean in 1996.

The planes instruments could measure the amount of fine specks, the cloud condensation nuclei. The instruments showed the expected conversion of dimethyl sulfide involving water vapor and ultraviolet light from the Sun, first into sulfur dioxide gas then into sulfuric acid vapor. The number of sulfuric acid molecules fluctuated quite a lot, but they remained far too low for them to club together, according to the prevailing theory.

Then, later in the flight, a detector on the aircraft encountered a great number of ultra-fine specks! In two minutes the count shot up from near zero to more than 30 million per liter of air. The number of free sulfuric acid molecules remained low.

That burst of ultra-fine specks should not have been there, with the available concentrations of sulfuric acid. This was a revelation of something quite important as it represented un-explained nucleation of the chief source of ccn over half of the globe.

To shorten the story here, it was recognized by Svensmark that ions created by cosmic rays could assist in making cloud condensation nuclei – hence clouds. The presence of electric charges would encourage the molecules to come together at lower concentrations of sulfuric acid vapor than would be possible without them. The ions would then stabilize the resulting embryonic specks while they assembled into larger specks. Subsequent calculations accounted for the results found on that research aircraft. Svensmark was able prove this!

Heavy electrons (muons) – liberated in the air by cosmic rays become attached to an oxygen molecule, a single electron is enough to make it attractive to water molecules. Several gather around making a water cluster. Activated by ozone and supplied by sulfur dioxide, the water cluster becomes a production center where sulfuric acid can be manufactured and accumulate.

The heavy electron (muon) is the glue still holding the cluster together. But when the cluster has stockpiled a few sulfuric acid molecules, and is still very small, it becomes stable on its own account. Then the electron (muon) can move on, find another oxygen molecule, and start instigating another cluster. So it acts as a catalyst continuing to amplify the important promoting chemistry for cloud formation.

The final summary, then goes with the following scenario. The cloud cover changes according to the changes in the Sun's activity (revealed by changes in the magnetic field induced by the solar wind and other effects) which regulates the number of cosmic rays that reach the Earth. The heavy electrons (muons) set free by the cosmic rays catalyze the clubbing together of sulfuric acid molecules, the most important source of cloud condensation nuclei.

While still open to more experiments, the chain of explanation from the stars to the clouds to the climate is now essentially complete. Please see "The Chilling Stars" by Svensmark and Calder.

CPSIA information can be obtained
at www.ICGtesting.com
Printed in the USA
BVHW010226290921
617736BV00002B/56